# NMR 32

Basic Principles and Progress

Editors:
P. Diehl E. Fluck H. Günther R. Kosfeld J. Seelig

Guest-Editor: B. Blümich

# Solid-State NMR III
# Organic Matter

Guest-Editor: B. Blümich

With contributions by
H. W. Beckham, F. Engelke, G. L. Hoatson,
W. E. J. R. Maas, D. Michel, H. W. Spiess,
W. S. Veeman, R. L. Vold

With 77 Figures and 8 Tables

Springer-Verlag
Berlin Heidelberg NewYork
London Paris Tokyo
Hong Kong Barcelona Budapest

ISBN-13:978-3-642-64751-2 e-ISBN-13:978-3-642-61223-7
DOI: 10.1007/978-3-642-61223-7


Softcover reprint of the hardcover 1st edition 1994

Library of Congress Catalog Card Number 93-9522



Typesetting: Thomson Press (India) Ltd, New Delhi
Offsetprinting: Saladruck, Berlin; Bookbinding: Lüderitz & Bauer, Berlin
SPIN: 10101492 51/3020 - 5 4 3 2 1 0 - Printed on acid-free paper

# Preface

Solid-State NMR is a branch of Nuclear Magnetic Resonance which is presently experiencing a phase of strongly increasing popularity. The most striking evidence is the large number of contributions from Solid-State Resonance at NMR meetings, approaching that of liquid-state resonance. Important progress can be observed in three areas: Methodological developments, applications to inorganic matter, and applications to organic matter. These developments are intented to be captured in three volumes in this series, each of them being devoted to more or less one of these areas.

The present volume on Solid-State NMR III is devoted mainly to organic matter. The recent developments of deuteron NMR and their applications are reviewed in the first chapter. Crosspolarization, MAS, and dynamic angle spinning are being explored for enhancement of information and sensitivity. In addition to the analysis of classical relaxation times and modern 2D spectra, detailed dynamic information becomes accessible from investigations of the relaxation time anisotropies. The second chapter examines cross-polarization in static and rotating solids under conditions of spin diffusion and thermal motion. The underlying dipole-dipole interaction is further exploited by the techniques described in the third chapter for studies of polymer-polymer miscibility. Short-range techniques are discriminated from long-range techniques based on spin diffusion. The use of these techniques is illustrated by a case study of PMMA/$PVF_2$ blends. The last chapter addresses novel methods and applications of two-dimensional exchange NMR for investigations of relative molecular orientations, polymer morphology, molecular dynamics, and macroscopic molecular order. Here the use of different mechanisms for frequency exchange by spin diffusion, molecular motion, and sample rotation is demonstrated with applications from polymer science.

The authors are thanked in particular for writing the contributions. Springer-Verlag has been very helpful in its assistance and editorial supervision.

B. Blümich

Aachen, May 1994 R. Kosfeld

## Guest-Editor


Prof. Dr. Bernhard Blümich
Lehrstuhl für Makromolekulare Chemie,
RTWH Aachen, Worringer Weg 1,
D-52056 Aachen, FRG

# Table of Contents

## Tables of Contents to Volumes 30 and 31

### Solid-State NMR I - Methods



### Solid-State NMR II - Inorganic Matter

# $^2$H-NMR Spectroscopy of Solids and Liquid Crystals

G. L. Hoatson and R. L. Vold[†]

Department of Physics and [†] Applied Science Program, College of William and Mary, P. O. Box 8795, Williamsburg, VA 23187-8795, USA

## Table of Contents



NMR Basic Principles and Progress, Vol. 32

Deuteron Magnetic Resonance spectroscopy ($^2$H-NMR) is an extremely powerful technique for investigating molecular structure and dynamics. This article covers the last five years (1988–93) and reviews the $^2$H-NMR literature on molecular crystals, disordered materials and surfaces, inclusion compounds, and thermotropic and lyotropic liquid crystals. The current status of conventional experiments, development of new techniques, and applications to representative systems are discussed, assessed and summarized.

One- or multi-dimensional spectra can be used to investigate structure and ordering. These provide information on the quadrupole interaction tensors (coupling constants and asymmetry parameters), orientational order parameters, and probability distribution functions. Dynamic information is obtained from $^2$H-NMR relaxation studies. The ultimate goal is to be able to determine the rate and trajectory of rotational and librational motion in the fast, intermediate and slow regimes. The kinetic windows accessed by different relaxation measurements are discussed, and strategies for combining experiments to obtain maximum information are emphasized. Finally, illustrative examples of $^2$H-NMR studies of a variety of physically and chemically interesting materials are presented. These are intended to demonstrate the scope and versatility of $^2$H-NMR for quantitative investigations of order and dynamics in solids and liquid crystals.

# 1 Introduction

Deuteron NMR spectroscopy is an extremely powerful technique for investigating molecular order and dynamics. In part, this is due to the fact that the dominant quadrupole interaction is both small enough to permit recording of undistorted spectra and sufficiently large to be a sensitive probe of local environment. Dipolar interactions and associated many body complications are usually negligible in $^{2}$H-NMR, which is a distinct advantage over solid state $^{13}$C and $^{1}$H NMR. As a result the spin dynamics of deuterons, in realistically complicated systems, is simple enough to permit rigorous density matrix analyses of complicated pulse sequences and relaxation phenomena. This greatly facilitates the development of new NMR techniques and allows critical evaluation of theoretical descriptions of order and dynamics.

This article describes new $^{2}$H-NMR techniques and their applications to studies of both structure and molecular motion in crystalline and disordered solids, surfaces, inclusion compounds, and thermotropic and lyotropic liquid crystals. It is focussed on papers published on these topics between January 1988 and August 1993. Many recent developments in $^{2}$H-NMR have originated from investigations of polymers and polymeric liquid crystals. While these techniques will be discussed in this article, their application to specific polymer systems has been recently reviewed [1, 2] and is therefore excluded.

This rest of this paper is organized in three sections: Section 2 is concerned with how information on structure and order can be obtained from $^{2}$H-NMR spectra. At the start of this part a nomenclature is established to describe the relevant Hamiltonians, principal axis systems, and quadrupolar splittings. While this may appear pedantic it is necessary because of the remarkable lack of uniformity in the literature. Section 3 will describe techniques for investigating and describing molecular dynamics in solids and liquid crystals. Finally, examples of how these methods have been applied to a diverse range of physically and chemically interesting systems will be reviewed in section 4. The quality of information obtained from these investigations demonstrates the power and utility of $^{2}$H-NMR as a probe of order and dynamics in solids and anisotropic fluids.

# 2 Structure

The motivation for acquiring $^{2}$H-NMR spectra of solids and liquid crystals is to determine the local electric environment of the deuteron and to use this to obtain information about molecular order and dynamics. The dominant quadrupolar interaction gives rise to an orientation dependent lineshape. For single crystals and uniaxial nematic liquid crystals (which orient spontaneously in a magnetic field) the spectrum consists of well resolved doublets for each chemically and magnetically distinct deuteron. The doublet splittings provide information on the relative orientation of the deuteron electric field gradient (EFG) tensors, and in liquid crystals, information about order parameters of the anisotropic fluid. Polycrystalline samples have an isotropic distribution of orientations with respect to the magnetic field, giving rise to the familiar powder pattern. For amorphous or partially ordered materials the $^2$H-NMR lineshape provides direct information about the orientational distribution function. In this section, the basic interactions responsible for the $^2$H-NMR spectrum are summarized, followed by a description of one and two dimensional techniques for obtaining information about structure and order.

## *2.1 Hamiltonians and Operator Formalism*

The time independent Hamiltonian operator for deuterons in a static magnetic field, $B_0$, is given by

$$H = H_Z + H_Q + H_\sigma + H_D \tag{1}$$

where the Zeeman Hamiltonian is:

$$H_Z = -\omega_0 I_Z \tag{2}$$

The Larmor frequency $\omega_0 = \gamma B_0$ is expressed in rad s$^{-1}$, and $I_Z$ is the operator for total $Z$ component of deuteron spin angular momentum. The quadrupolar ($H_Q$), dipolar ($H_D$) and chemical shift ($H_\sigma$) terms may all be regarded as first order perturbations of the Zeeman Hamiltonian. Each of these interactions can be represented by a diagonal, time independent second rank tensor in an appropriate principal axis system (PAS). For example, the dipolar interaction between two deuterons in a $CD_2$ group has the principal Z-axis along the D-D internuclear vector, while for the quadrupole interaction of each deuteron, the principal Z-axis is along the local electric field gradient tensor. Generally this is parallel to the C—D bond [3]. The observed spectra depend on the orientation of the PAS with respect to laboratory fixed axes (LAB), and a unified description of their combined orientation dependence is best developed using irreducible spherical tensor notation.

### *2.1.1 Quadrupole Interactions*

For an isolated deuteron the quadrupole Hamiltonian, $H_Q$, is given by

$$H_Q = \sum_{m=-2}^{2} (-1)^m A_{-m}^{(2,Q)} T_m^{(2,Q)} \tag{3}$$

where the spin operators $A_m$ are defined as

$$A_0^{(2,Q)} = \frac{4}{\sqrt{6}}(3I_Z^2 - I \cdot I) \tag{4a}$$

$$A^{(2,Q)}_{\pm 1} = \mp \tfrac{1}{2}(I_Z I_{\pm} + I_{\pm} I_Z) \tag{4b}$$

$$A^{(2,Q)}_{\pm 2} = \tfrac{1}{2} I^2_{\pm} \tag{4c}$$

The lattice variables are given by

$$T^{(2,Q)}_{m}(\mathrm{LAB}) = \sum_{k=-2}^{2} D^{(2)*}_{mk}(\Omega_{LP}) T^{(2,Q)}_{k}(\mathrm{PAS}) \tag{5a}$$

where

$$T^{(2,Q)}_{0}(\mathrm{PAS}) = \frac{\sqrt{6}}{4}(e^2 q_{ZZ} Q/\hbar) \tag{5b}$$

$$T^{(2,Q)}_{\pm 1}(\mathrm{PAS}) = 0 \tag{5c}$$

$$T^{(2,Q)}_{\pm 2}(\mathrm{PAS}) = \tfrac{1}{4}\eta (e^2 q_{ZZ} Q/\hbar) \tag{5d}$$

The quadrupole coupling constant, $(e^2 q_{zz} Q/\hbar)$, is expressed in rad s$^{-1}$ in Eq. (5), and the asymmetry parameter $\eta$ of the electric field gradient tensor is defined as

$$\eta = (q_{xx} - q_{yy})/q_{zz} \tag{5e}$$

Equation (5a) relates tensor components in the laboratory fixed (L) coordinate system to those in the principal axis (P) system of the electric field gradient tensor, by means of Wigner rotation matrix elements, $D^{(2)}_{mk}(\Omega_{LP})$ [4], with Euler angle $\Omega_{LP} = (\alpha_{LP}, \beta_{LP}, \gamma_{LP})$. It is often necessary to describe this single transformation by a series of rotations. If a rotation from the PAS to a molecule fixed frame (M) is defined by $\Omega_{MP}$ and a subsequent rotation to the laboratory fixed frame by $\Omega_{LM}$, then

$$D^{(2)}_{mk}(\Omega_{LP}) = \sum_{l=-2}^{2} D^{(2)}_{ml}(\Omega_{LM}) D^{(2)}_{lk}(\Omega_{MP}) \tag{6}$$

The combined Hamiltonian, $H^0 = H_Z + H_Q$ is obtained by retaining only the secular part of $H_Q$, i.e., that part which commutes with $H_Z$. Making use of the general relation

$$[I_Z, A_m] = m A_m \tag{7}$$

it is found that

$$H^0 = H_Z + H_Q = -\omega_0 I_Z + \tfrac{1}{3}\omega_Q (3I_Z^2 - I \cdot I) \tag{8}$$

The energy level diagram and allowed transitions for this Hamiltonian are shown in Fig. 1.

The quadrupole term lifts the twofold degeneracy of transitions between the Zeeman eigenstates at the Larmor frequency, $\nu_0$, and results in two transitions at frequencies, $\nu_0 \pm \nu_Q$, with

$$2\nu_Q = \frac{2\omega_Q}{2\pi} = \tfrac{3}{4}(e^2 q_{zz} Q/h)[(3\cos^2\beta_{LP} - 1) + \eta \sin^2\beta_{LP} \cos 2\gamma_{LP}] \tag{9}$$

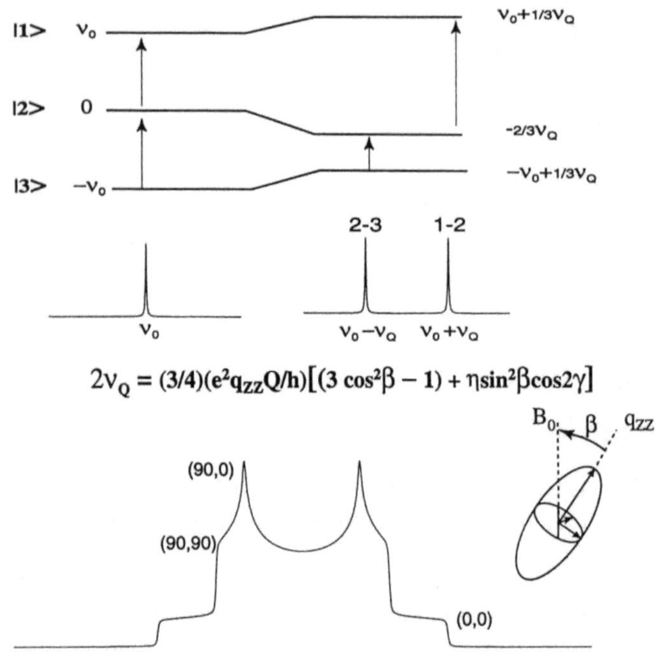


**Fig. 1.** The energy level diagram for a spin-1 deuteron in a magnetic field. *On the left* the allowed transitions are indicated and the single line NMR spectrum is sketched. *On the right*, the energies of the eigenstates are modified by the first order quadrupole interaction. Now the allowed transitions give a two line spectrum, with separation of $2\nu_Q$. In the *lower half* of the figure the expression for the quadrupolar splitting is given, which indicates its dependence on the orientation $(\beta, \gamma)$ of the static field and the electric field gradient tensor. An isotropic distribution of orientations results in the familiar powder pattern, which is shown in the *bottom trace*. Next to the discontinuities in the line-shape are the corresponding $(\beta, \gamma)$ values

For a single deuteron with the principal Z-axis of the EFG tensor along the magnetic field direction ($\beta_{LP} = 0$), the spectrum consists of two sharp lines, shown schematically in Fig. 1. For a polycrystalline powder, which has an isotropic distribution of angles $\beta_{LP}$ and $\gamma_{LP}$, the spectrum is a superposition of doublets and the overall lineshape is the characteristic powder pattern. As shown at the bottom of Fig. 1, three singularities occur in the spectrum corresponding to specific orientations: for $\beta_{LP} = 0^0$ at $\pm\frac{3}{4}(e^2q_{zz}Q/h)$, for ($\beta_{LP} = 90°$, $\gamma_{LP} = 0$) at $\mp\frac{3}{8}(e^2q_{zz}Q/h)(1+\eta)$, and for ($\beta_{LP} = 90°$, $\gamma_{LP} = 90°$) at $\mp\frac{3}{8}(e^2q_{zz}Q/h)(1-\eta)$.

Typical values for deuteron quadrupole coupling constants in rigid C—D bonds depend on the hybridization of the carbon and range from about 150 to 200 kHz, and they are somewhat larger for N—D and O—D moieties. The largest known deuteron QCC is 1.41 MHz in $D_2$ [5], but this is exceptional. It is important to realize that while the intrinsic electronic asymmetry parameter for a deuteron in a C—D bond is usually small ($\eta \leq 0.1$), no such restriction is expected for electric field gradient tensors which are averaged over rapid, asymmetric molecular motion. In fact, the measurement of motionally averaged quadrupole coupling parameters (combined with accurate estimates of the static values) is a widely used source of information on molecular dynamics, and will be discussed in detail in Sect. 3.4.

### *2.1.2 Dipolar Interactions*

The dipolar Hamiltonian for a deuteron (spin I) coupled to another spin ($S$) is given by an equation similar to that for the quadrupole Hamiltonian:

$$H_D = \sum_{m=-2}^{2} (-1)^m A_{-m}^{(2,D)} T_m^{(2,D)} \tag{10}$$

Now the spin operators are defined as

$$A_0^{(2,D)} = \frac{1}{\sqrt{6}}(3I_Z S_Z - I \cdot S) \tag{11a}$$

$$A_{\pm 1}^{(2,D)} = \mp \tfrac{1}{2}(I_Z S_\pm + I_\pm S_Z) \tag{11b}$$

$$A_{\pm 2}^{(2,D)} = \tfrac{1}{2} I_\pm S_\pm \tag{11c}$$

and the lattice variables are given by

$$T_m^{(2,D)}(\mathrm{LAB}) = \sum_{k=-2}^{2} D_{mk}^{(2)*}(\Omega_{LP}) T_k^{(2,D)}(\mathrm{PAS}) \tag{12a}$$

where

$$T_0^{(2,D)}(\mathrm{PAS}) = -\sqrt{6}\hbar\gamma_I\gamma_S r_{IS}^{-3} \tag{12b}$$

$$T_{\pm 1}^{(2,D)}(\mathrm{PAS}) = T_{\pm 2}^{(2,D)}(\mathrm{PAS}) = 0 \tag{12c}$$

and $r_{IS}$ is the internuclear distance between spins $I$ and $S$. As for the quadrupolar interaction, only the secular part of the dipolar Hamiltonian is needed to calculate the energy eigenvalues and predict the static spectrum. There are two distinct types of dipolar interactions, homonuclear and heteronuclear. In the homonuclear case the complete operator $A_0^{(2,D)}$ commutes with the zero order Zeeman Hamiltonian, $H_Z = -\omega_0(I_Z + S_Z)$. In the heteronuclear case the zero order term is ($H_Z = -\omega_I I_Z - \omega_S S_Z$) and the secular part of $A_0^{(2,D)}$ reduces to $2/\sqrt{6} I_Z S_Z$. Eigenfunctions, transition frequencies, and intensities for deuterons experiencing homonuclear and heteronuclear dipolar couplings are tabulated in the literature [6].

In order to assess the relative magnitude of dipolar and quadrupolar contributions, it is useful to examine the ratio $T_0^{(2,D)}(\mathrm{PAS})/T_0^{(2,Q)}(\mathrm{PAS}) = 4\,\mathrm{h}\,\gamma_I\gamma_S r_{IS}^{-3}/(e^2 q_{zz} Q/h)$. For methylene deuterons, $CD_2$, typical values are $r_{IS} = 1.7$ Å and QCC = 169 kHz, and this ratio is only 0.013. Thus, while they may contribute to the underlying homogeneous linewidth, deuteron-deuteron dipolar interactions are rarely resolved in powder spectra [7]. However, many of the modern techniques described in this review are aimed at improving spectral resolution and information content by eliminating the dominant quadrupolar interaction. In these experiments $^2H$-$^2H$ and $^1H$-$^2H$ dipolar couplings, second order quadrupole effects, and chemical shift anisotropy will play increasingly important roles as sources of information on structure and dynamics.

In spectra of single crystals and monodomain liquid crystals dipolar coupling patterns from $D_2$ and $D_3$ groups are often sufficiently well resolved to provide reasonably accurate experimental values of the dipolar coupling constants. However, dipole coupling among more distant deuterons is usually manifested only as line broadening. Heteronuclear proton-deuteron dipolar couplings, which are larger by a factor $(\gamma_H/\gamma_D) = 6.51$, are more useful in structural studies. An extensive discussion of deuteron dipolar coupling of molecules in liquid crystal solvents has been given by Emsley and Lindon [6].

### *2.1.3 Chemical Shift Interactions*

A complete discussion of the chemical shift Hamiltonian and its transformations under rotation has been given by Haeberlen [8]. To summarize, the secular part of $H_\sigma$ is given by:

$$H_\sigma^0 = -\omega_0 I_Z\{\sigma_{iso} + \tfrac{1}{2}\delta(3\cos^2\beta - 1 + \eta\sin^2\beta\cos 2\gamma)\} \tag{13}$$

Here the trace of the shielding tensor is $\sigma_{iso} = \frac{1}{3}(\sigma_{xx} + \sigma_{yy} + \sigma_{zz})$, the chemical shift anisotropy is defined as $\Delta\sigma = \sigma_{zz} - 1/2(\sigma_{yy} + \sigma_{xx}) = 3/2\,\delta$, and the asymmetry parameter of the shielding tensor is defined as $\eta = (\sigma_{xx} - \sigma_{yy})/\sigma_{zz}$ [8].

For protons in diamagnetic materials $\Delta\sigma$ is rarely larger than 10–15 ppm, and this implies that effects of chemical shift anisotropy for deuterons are usually small enough to ignore. However, Hamiltonians of the same form as Eq. (13), but with significantly larger values for $\delta$ and $\sigma$, can arise in paramagnetic systems [9–11].

### *2.1.4 Notation*

In this paper, the symbol $\nu$ is reserved for frequencies expressed in Hertz and $\omega$ for frequencies in rad s$^{-1}$. Unfortunately, there is little notational consistency in the literature, and both $2\nu_Q$ and $\nu_Q$ are referred to as the quadrupole splitting without reference to units, orientation angles, or quadrupole coupling constants (QCC). Occasionally the separation between the two parallel edges ($\beta = 0^0$) of the powder pattern is incorrectly called the QCC. In subsequent section of this paper literature results will be converted, if necessary, to the notation established by Eqs. (1) to (9). The quadrupolar splitting, $2\nu_Q$, is defined as the frequency difference between the doublet components and QCC will be used to refer to the quadrupole coupling constant, $(e^2q_{ZZ}Q/h)$, expressed in Hertz. A uniform convention with regard to labeling of principal axes and magnitudes of tensor components will be adopted [12, 13], such that, $|T_{zz}| \geqq |T_{yy}| \geqq |T_{xx}|$ and $\eta = (T_{xx} - T_{yy})/T_{zz}$ with $0 \leqq \eta \leqq 1$. This convention is usually followed in articles on quadrupolar nuclei, but the reader is warned that in Haeberlen's monograph [8], and much of the subsequent chemical shift literature, the definitions of principal $x$ and $y$ axes for the shielding tensor have been interchanged. Thus according to Ref. [8], Eq. (13) is valid as written but $x$ and $y$ are labeled such that $|T_{zz}| \geqq |T_{xx}| \geqq |T_{yy}|$ and $\eta$ is defined as $(\sigma_{xx} - \sigma_{yy})/\sigma_{zz}$. This confusion has led some authors to retain the eigenvalue convention $|T_{zz}| \geqq |T_{yy}| \geqq |T_{xx}|$, but to alter the definition of the asymmetry parameter, $\eta = (\sigma_{yy} - \sigma_{xx})/\sigma_{zz}$. This necessitates changing the sign of the term with $\eta$ in Eq. (13). A proposal for standard conventions for shielding tensors has been presented by Mason [14].

## 2.2 *Quadrupole Echo*

Recording undistorted solid state $^2$H spectra places severe demands on conventional instrumentation for Fourier transform NMR. The full width of a $^2$H powder pattern can be on the order of 300 kHz, and a fast digitizer ($\sim$1 μs dwell time) is needed to obtain spectra with well defined baselines. A more difficult problem is posed by the need for high pulse power ($\sim$1.6 kW) to achieve uniform excitation across the lineshape. This conflicts with the necessity of obtaining data immediately, or at most a few microseconds, after the pulse. With current instrumentation it is possible, in a 5 mm sample coil, to obtain 1.5–2.0 μs 90° pulses and to record signal about 15–20 μs after the pulse. The standard procedure [15–17] for avoiding distortions due to this dead time is to use a Quadrupolar Echo pulse sequence (QE), $90_x°$-$\tau$-$90_y°$, and record the echo signal occurring at time $\tau$ after the second pulse. Bloom et al. [18] calculated distortions introduced by finite amplitude pulses; reduced intensity for large $\nu_Q$ and time shift of the echo maximum. Griffin [19] and Henrichs et al. [20] have described procedures for dealing with additional distortions, such as the unwanted virtual FID produced by the first pulse in the sequence. It is necessary to phase cycle both pulses not only to avoid receiver channel imbalance but also undesired interference from unwanted coherences. All these procedures are now quite standard, and the quadrupole echo technique is currently the method of choice for obtaining $^2$H-NMR spectra.

In spite of the success of the simple quadrupole echo sequence, it should be remembered that two potentially important sources of spectral distortion are present. Firstly, the echo decay is a faithful image of the true FID only to the extent that the $\delta$-function rf pulses result in complete time reversal of the Hamiltonian. This requires that transverse relaxation occurring in the time, $\tau$, between the pulses is negligible. For samples like oriented liquid crystals which have $T_2$ of a few milliseconds this presents no problem. However, in many cases molecular motion leads to orientation dependent $T_2$s on the order of 10–100 μs; then the dependence of quadrupole echo lineshape on pulse spacing becomes a powerful tool for measuring rates and mechanisms of rotational processes [21]. Lineshape analysis is a well established procedure for probing molecular dynamics and recent work exploiting this will be reviewed in Sect. 3.4.

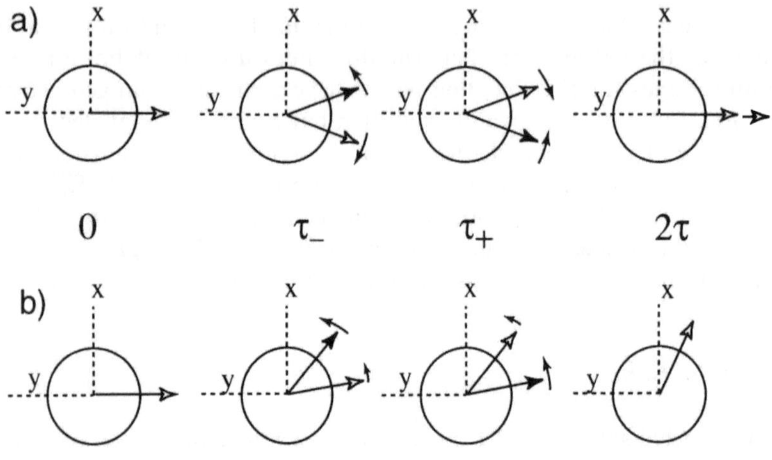


**Fig. 2.** This depicts the time dependence of the magnetization vectors during a quadrupolar echo sequence which begins with a $90_x$ pulse. The *top series* shows the on resonance case, where the fast (*closed arrow*) and slow (*open arrow*) coherences are refocused perfectly by a $90_y$ pulse to give an echo at $2\tau$. The *lower series* shows the off resonance case. Now the rotating frame is precessing at frequency $\delta$, and the phase of the second 90° pulse must be adjusted accordingly. The phase of the echo signal is then $2\delta\tau$ instead of zero

A second source of spectral distortion in the ordinary QE sequence is it's inability to compensate for errors introduced by evolution under the chemical shift Hamiltonian. These are analogous, but not identical, to errors which arise if the pulses are applied off resonance. When there is no offset then the x-component magnetization of every counter rotating doublet remains zero at all times after an ideal $90_x$° pulse, and the $90_y$° pulse completely refocuses the quadrupolar interaction (see Fig. 2a). A resonance offset, $\delta$, leads to precession of the antiphase component of the quadrupolar doublet by $\delta\tau$ during the time between pulses, as shown in Fig. 2b. An echo is still obtained if the phase of the second pulse is set to $y + \delta\tau$. However, if there are two deuterons with different chemical shifts, the phase of the second pulse cannot be adjusted to compensate for both.

Distortions due to different chemical shifts can be avoided by interleaving 180° pulses in the sequence at appropriate times [22]. Siminovitch et al. [10] showed that this method works for small, orientation dependent chemical shifts in paramagnetically spin labeled membranes, and Wittebort [11] showed that the same technique is successful for the much larger shifts found in paramagnetic solids.

### 2.3 Composite Pulse Techniques

It could be argued that the problem of pulse power and adequate spectral bandwidth has been solved for routine deuteron spectroscopy of solids, since

1.5 μs 90° pulses correspond to a sinc function excitation profile of width ±333 kHz. However, such intense rf fields ($B_1 = 156$ kHz) are difficult, if not impossible, to achieve with sample coils larger than 5 mm and for Larmor frequencies much above 50 MHz. Furthermore, even if the spectral coverage of the (1.5 μs) 90° pulse is adequate, that of the corresponding (3.0 μs) 180° pulse is often not. Such considerations provided strong motivation to develop more efficient ways of using the available pulse power. Levitt et al. [23] first suggested composite 90° pulses to reduce distortions in quadrupole echo lineshapes. Siminovitch et al. [24], Heaton et al. [25], and Raleigh et al. [26] extended these ideas to more complex families of sequences designed to achieve uniform spectral inversion. An important advance was made by Wimperis, who described an iterative procedure for designing both frequency selective and highly uniform (broadband) pulse sequences for inversion and for excitation of J-order in coupled spin-1/2 systems and quadrupole order for spin-1 [27]. Hoatson demonstrated that one of these broadband sequences was capable of exciting uniform quadrupolar order across a full width deuteron powder pattern, and that echo detection gave distortion free spectra [28]. Callahan et al. have investigated broadband sequences computationally and have designed optimized composite pulse sequences for inversion and excitation of quadrupolar order [29].

There appear to be many advantages to composite pulse sequences and no serious drawbacks to their routine implementation. The importance of being able to design rf pulse sequences which generate particular states of the density matrix will become more obvious in Sect. 3 which discusses the measurement of specific deuteron relaxation times.

## 2.4 *Hadamard Spectroscopy*

Hadamard NMR provides an interesting alternative to the single minded quest for still more pulse power. This technique has been successfully applied to deuteron NMR in solids and liquid crystals [30]. It may prove to be the only feasible approach to obtaining wideline spectra from quadrupolar nuclei with large QCCs ($^{27}$Al, $^{17}$O and $^{14}$N) where there is no hope of achieving adequate excitation bandwidth using current amplifier technology.

Hadamard spectroscopy is a form of stochastic excitation [31, 32] which relies on a low power pseudo random pulse sequence to excite the linear response of the spin system. The main advantage over QE excitation is the reduction in pulse power, which can be four orders of magnitude [30, 33]. Because the linear response is required, the rf pulses are made short ($t_p \sim 200$ ns) to ensure small flip angles (typically 0.05°–2.0°). Thus the bandwidth of the pulsed excitation can be several MHz and this is no longer a major limitation. In addition, there is the possibility of recording spectra of systems where the molecular motion is in the intermediate time regime and conventional pulsed excitation fails or where the longitudinal relaxation times, $T_{1Z}$, are very long and adequate signal averaging is prohibitive.

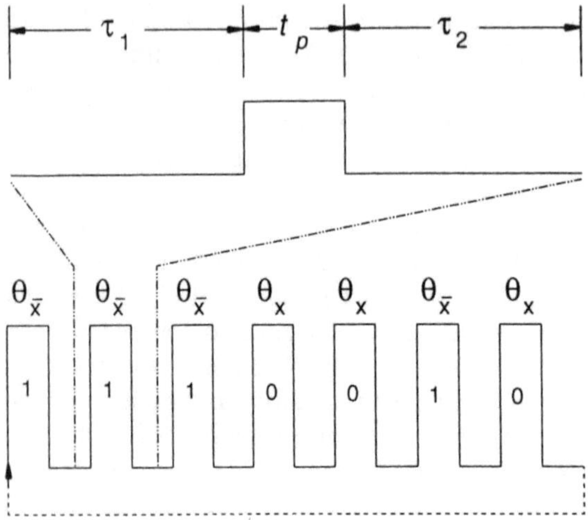


**Fig. 3.** The low power pulse sequence for Hadamard NMR. The phases of the pulses are modulated pseudorandomly by a maximum length binary sequence (mlbs). This example is for seven pulses; *1* denotes $-x$ phase and *0* $+x$. The pulse sequence is repeated continuously and one data point is acquired between the finite length pulses ($t_p$). The dwell time is thus ($\tau_1+\tau_2+t_p$)

In Hadamard spectroscopy the rf excitation is applied in a pseudo continuous fashion with random modulation of the phase or amplitude of the pulses. In the original deuteron experiments [30] a series of equally spaced rf pulses were phase modulated ($\pm x$) according to ($2^m$-1, integer m) values of a maximum length binary sequence (mlbs) which was repeated cyclically. This is shown in Fig. 3 for the seven pulse sequence, m = 3. The excitation and detection of the response were performed in a time shared manner. One complex data point was acquired in the time window between successive pulses and thus the dwell time is ($\tau_1 + t_p + \tau_2$). The conventional FID signal is derived by cross correlation of the rf excitation with the magnetization response, and subsequent Fourier transformation yields the NMR spectrum. In the linear regime, when the pseudo random modulation of the excitation is a mlbs, the cross correlation is accomplished by a Hadamard transformation.

The dramatic decrease in power requirement was demonstrated by a S/N comparison on HMB, using a constant acquisition time (~60 s) [33]. The results are reproduced in Fig. 4, which shows (a) the Hadamard spectrum and (b) the QE Fourier spectrum. These were collected with identical filter bandwidth (±500 kHz) but radically different power levels; 0.1 W and 1 kW, respectively. The spectrum shown in (c) is a Fourier spectrum taken with a matched filter (±30 kHz). The filters lead to an obvious S/N improvement in the Fourier spectrum but they cannot be used in Hadamard NMR because they introduce a time delay and receiver ringing, both of which interfere with sampling of the stochastic response immediately after each excitation pulse. One alternative currently being explored is the use of a crossed coil probe design to eliminate the time shift between excitation and detection of the response [34]. This should remove the current bandwidth limitation (±53 kHz), as well as baseline distortions and first order phase corrections in the FT spectrum.

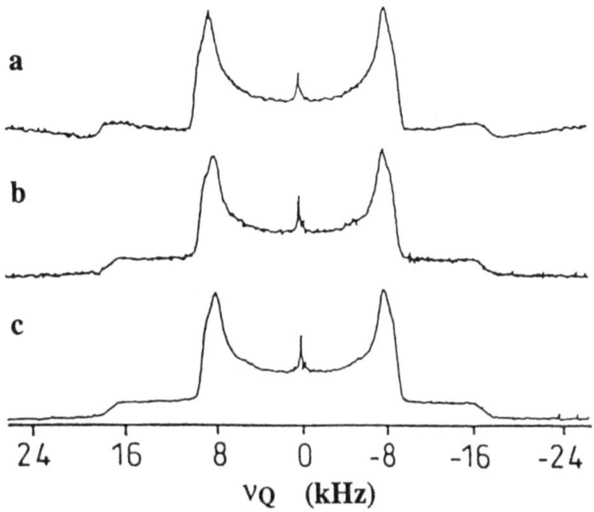


**Fig. 4.** $^2$H-NMR spectra of HMB all collected in approximately 60s. (*a*) The stochastic response was acquired using 1023 pulses, Hadamard and then Fourier transformed. (*b*) This is the Fourier spectrum obtained with the quadrupolar echo sequence. For both (*a*) and (*b*) the filters were set to ±500 kHz, but the pulse power was $10^4$ lower in (*a*). (*c*) Is the Fourier spectrum with matched filters, ±30 kHz. Reproduced with permission from Greferath et al. [33]

Studies on saturation effects demonstrate that too much power leads to systematic, noise like features on the powder lineshape. This is a direct result of violating the linearity condition. Failure to observe the cyclicity of the sequence, by including a recycle delay after each ($2^m$-1) pulses, gives high frequency noise which can be effectively suppressed using exponential multiplication. The spectra seem insensitive to finite pulse widths, and thus the duty cycle of the mlbs excitation. These three observations have profound implications for the implementation of this technique on commercial spectrometer systems and hence its wide spread use.

Improvement of noise excitation schemes and development of the appropriate cross correlation functions will permit detection of the nonlinear response of the spin system. Recording of these higher order responses is of interest because it is potentially a very efficient method of obtaining multi dimensional $^2$H-NMR spectra.

## 2.5 *Magic Angle Spinning*

There is a growing realization that magic angle spinning (MAS) may offer an attractive alternative to the ubiquitous quadrupole echo for routine observation of solid state $^2$H-NMR spectra. First demonstrated experimentally for deuterons by Ackerman, Eckman and Pines [35], this technique consists of applying a single pulse and recording the signal from a sample rotating at the magic angle

($\beta = \cos^{-1}(3^{-1/2})$) with respect to the external magnetic field. This consists of a train of rotational echoes [36] whose tips trace out a decaying signal which is independent of the first order quadrupole interaction. By synchronizing data acquisition with the spinning speed and acquiring only the echo maxima, Pines and coworkers [37–42] exploited this suppression of the quadrupole interaction to obtain resolved chemical shifts. Fourier transformation of the entire echo train yields a spectrum which is broken up into spinning sidebands, whose envelope yields an approximation to the original powder pattern lineshape. Schadt et al. [43] and Kristensen et al. [44, 45] have pointed out that the line narrowing achieved by deuteron MAS results in a S/N gain over the static QE spectrum. An especially impressive example of this sensitivity increase in one dimensional deuteron MAS is the natural abundance spectrum of dimethylsulfone reported by Reichert et al. [46]. This spectrum, reproduced in Fig. 5, was obtained with only 17 minutes signal averaging.

Kustanovitch et al. [47] showed that off axis spinning (VAS) scales the quadrupole interaction and can be used to enhance the sensitivity of the powder lineshape to effects of molecular motion. The spectra obtained are reminiscent of temperature dependent lD lineshape studies, since it is the scaled quadrupolar interaction that must be compared with the correlation time to determine which motional regime is appropriate. Density matrix treatments of the combined effects of sample spinning and molecular motion on deuteron lineshapes have been presented by Kristensen et al. [44] and by Duer and Levitt [48]. These calculations are based on the stochastic Liouville equation and are claimed to be less cumbersome than methods based on Floquet formalism [49, 50].

A simple, yet very powerful, variation of magic angle spinning is provided by Two dimensional One Pulse Spectroscopy (TOPS) [51, 52]. In this magic angle spinning technique the signal following a single pulse is acquired. A two dimensional data set is subsequently constructed by choosing N points starting at the top of each rotational echo to define the $t_2$ domain, and defining increments in $t_1$ as integral multiples of the rotor period. It is convenient, though

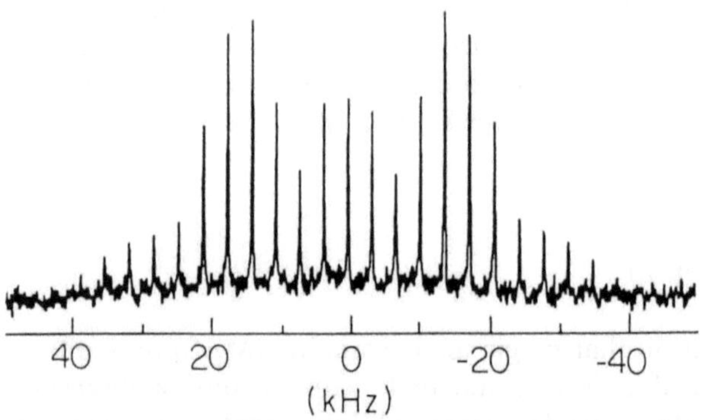


**Fig. 5.** The natural abundance deuteron MAS spectrum of dimethylsufone. This spectrum was collected in 17 min, spinning at $\nu_r = 3.55$ kHz. Reproduced with permission from Reichert et al. [46]

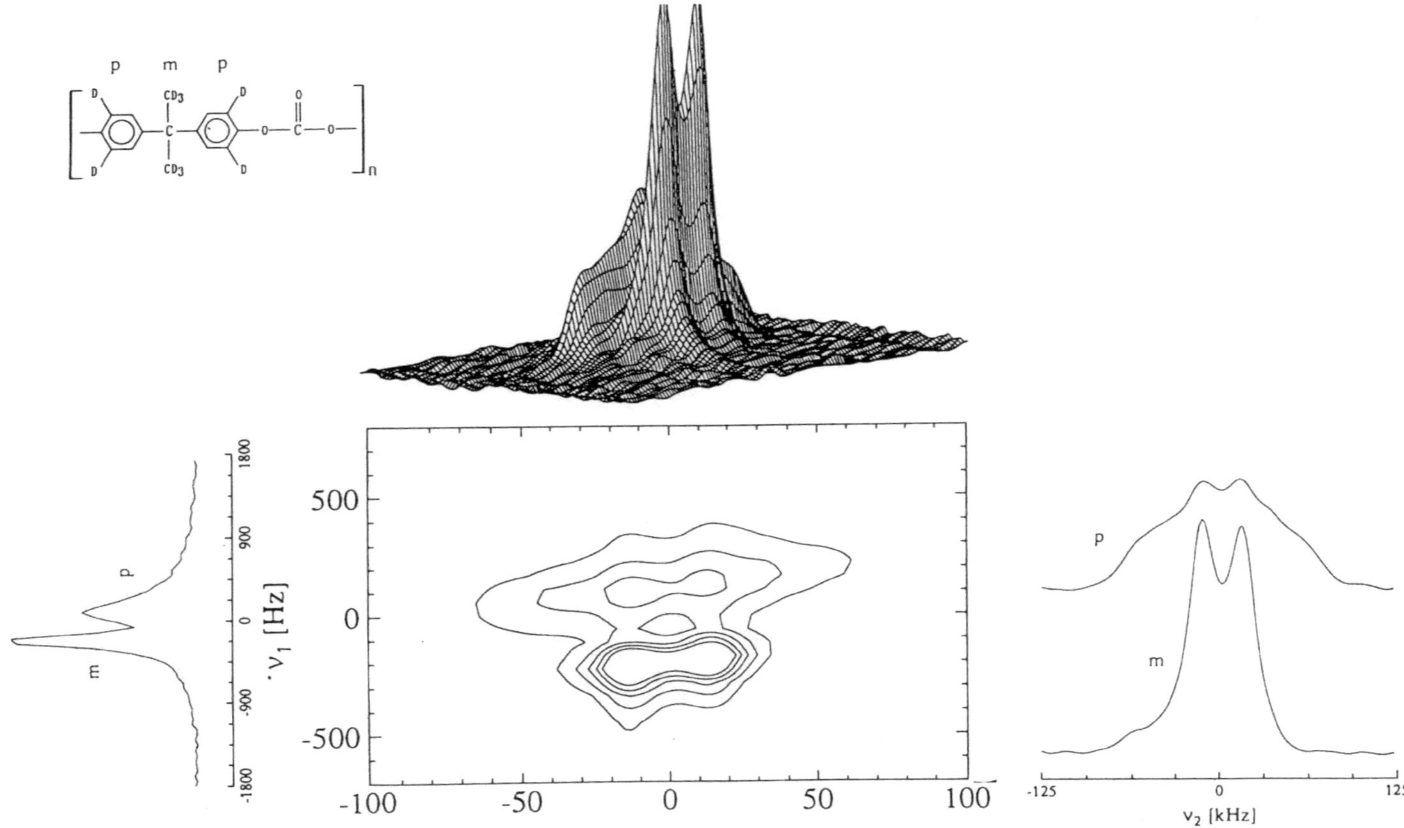


**Fig. 6.** The two dimensional one pulse (TOPS) spectrum of bisphenol-A polycarbonate. Both the stack and contour plots of the data are shown. Along the isotropic dimension $\nu_1$ (obtained from the rotary echo maxima) is the chemical shift spectrum showing resolved methyl (*m*) and phenyl (*p*) resonances. Shown on the *right* are slices (parallel to $\nu_2$) through the methyl and phenyl peaks; this separates the two anisotropic quadrupole powder patterns. Reproduced with permission from Bluemich et al. [51]

not essential, to adjust the spinning speed and dwell time so that a point is acquired exactly at the top of each rotational echo. Double Fourier transformation of the reconstructed data set yields an isotropic chemical shift spectrum along the $\nu_1$ axis, and the corresponding quadrupole powder pattern along the $\nu_2$ dimension. This technique is illustrated for bisphenol-A polycarbonate in Fig. 6, where the aromatic and methyl deuterons were resolved [51]. TOPS has great potential in high magnetic fields and for double quantum spectroscopy.

When calculating a NMR spectrum, dropping the nonsecular parts of the Hamiltonian amounts to ignoring all off diagonal elements of $H_Q$, $H_D$, and $H_\sigma$. This first order approximation is adequate for most purposes, but important exceptions arise in experiments (like MAS) which are designed to suppress the dominant first order quadrupolar interaction. It can be shown [53] that second order corrections to the energy levels are eigenvalues of the Hamiltonian operator $H^{(2)}$, given by

$$H^{(2)} = \sum_{m \neq 0} \left(\frac{1}{m}\right) T^{(2)}_{m} T^{(2)}_{-m} [A^{(2)}_{-m}, A^{(2)}_{m}] \tag{14}$$

Few explicit calculations of second order effects in deuteron NMR have appeared in the literature. Perhaps the most important of these is the second order quadrupole interaction, which cannot be removed by spinning the sample at any one angle. The order of magnitude of the second order quadrupole shift is [54] $\delta^{(2)} = \nu_Q^2/24\nu_0$; at 46 MHz this is $\sim$1 ppm which is comparable in magnitude to deuteron chemical shift difference. In addition, homonuclear $^2$H-$^2$H dipolar coupling is only partially suppressed by spinning and may contribute to the residual linewidth. Calculation of these effects is not entirely trivial. Ignoring relaxation and molecular motion, the proper procedure [53] is to transform Eq. (14) into a rotor fixed frame, average the result over a single rotor period, and finally sum over the Euler angles relating the rotor fixed axes to the deuteron PAS.

Schmidt et al. have recently described a powerful new way of using $^2$H rotational echoes to obtain highly accurate, selective measurements of internuclear distances up to ca. 5 Å [55]. Selectivity is achieved by doubly labeling the molecule of interest with $^2$H and another spin, S, such as $^{13}$C or $^{15}$N. The $^2$H rotational echo train is then recorded, with and without 180° pulses applied to the non resonant S-spins. Simulation of the resulting difference spectra gives the desired internuclear distance, in favorable cases with accuracy approaching 0.1 Å. The acronym REDOR (Rotational Echo Double Resonance) for this experiment is slightly misleading; since it is usually necessary to decouple protons this is actually a triple resonance experiment.

## 2.6 Cross Polarization

According to the general theory of cross polarization, it should be possible to transfer polarization from protons to deuterons with a gain in sensitivity given by [42]

$$(S^{CP}/S^0) = \gamma_H \gamma_D (1 + 8 n_D/n_H)^{-1} \tag{15}$$

Here $S^{CP}$ and $S^0$ are the deuteron signals with and without cross polarization, respectively, and it is assumed that $n_D$ deuterons are dipolar coupled to $n_H$ protons. Spin relaxation is ignored in the derivation of Eq. (15), as are practical problems related to the difficulty of matching the spin locking levels of the protons and deuterons. Pines and coworkers [42, 56] have considered these questions both theoretically and experimentally. They showed that the expected enhancements, on the order of 2–5, can be obtained for both the single and double quantum transitions, with and without magic angle spinning.

Given these promising early results, it seems surprising that cross polarization is rarely used in solid state deuteron NMR. In part, this is due to technical difficulties associated with achieving adequate spectral coverage of the deuterons while simultaneously decoupling protons (without destroying the sample). More fundamentally, enhancement by cross polarization decreases rapidly as the ratio $n_D/n_H$ of rare to abundant spins increases. Thus, increasing the level of deuteration is generally more effective at increasing sensitivity than is cross polarization of isotopically dilute samples.

Cross polarization techniques are invaluable in studies of interfaces and phase separation in polymer blends. If one component of a heterogeneous mixture is deuterated and the other components are unlabelled, transfer of polarization from protons to deuterons yields a $^2$H-NMR signal originating from the interfacial region. For these types of experiments it is the spatial specificity that is the crucial feature. Zumbalydis [57] showed that since proton to deuteron cross polarization is not very efficient, the increased sensitivity provided by magic angle spinning is necessary to achieve adequate S/N for detection.

## 2.7 *Zero Field NMR*

In a zero field NMR experiment the sample is shuttled from a large polarizing field to a collinear, intermediate field which is then switched off suddenly. The spin system then evolves under the zero field Hamiltonian for a time $t_1$. The intermediate field is switched back on, the sample is returned to high field, and the signal is acquired during $t_2$ using multiple echoes to improve sensitivity [58–62]. The zero field spectrum can be obtained by two dimensional data analysis procedures. A one dimensional version of this experiment consists of recording the integral of the echo train as a function of evolution time $t_1$, followed by Fourier transformation. Use of the intermediate field preserves the polarization while easing the task of rapid field cycling, and this constitutes an important technical advance over the early field cycling procedures [63, 64].

Obviously the deuteron zero field spectrum is dominated by the quadrupole interaction. However, since there is no external field, spectra from different

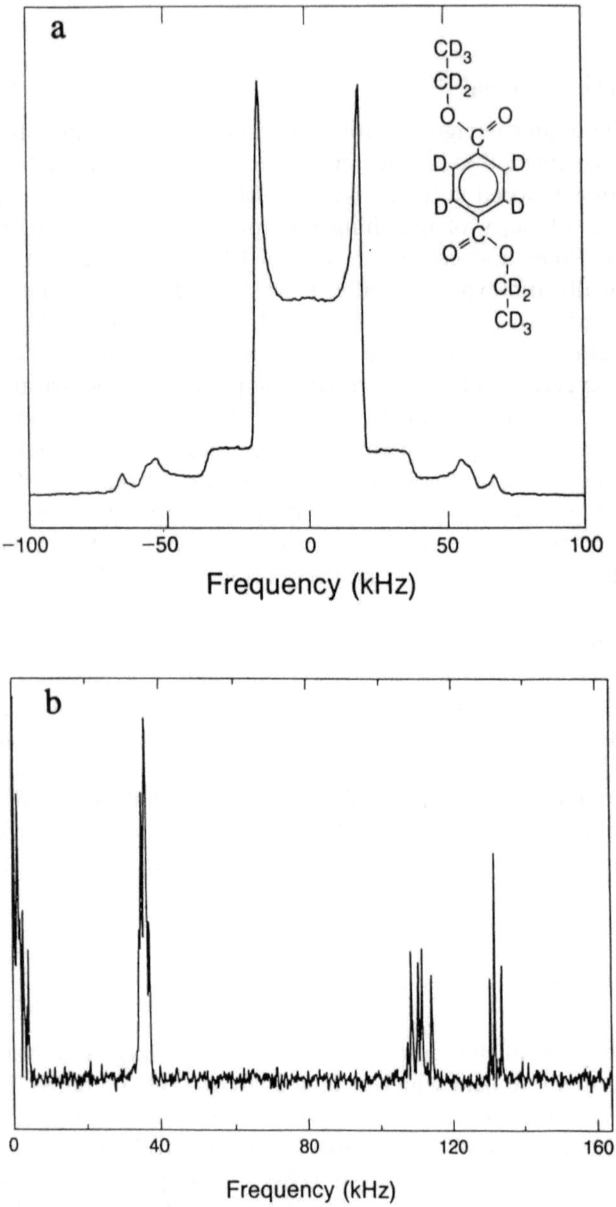


**Fig. 7a.** The high field quadrupolar echo spectrum of perdeuterated diethylterephalate (DETP) shows singularities from three overlapping powder patterns. **b** Zero field NQR spectrum of the same sample. From this data quadrupolar coupling constants and asymmetry parameters were established for five inequivalent deuterons. Adapted with permission from Millar et al. [60]

crystallites in a powder have the same transition frequencies and intensities regardless of orientation. The method is therefore ideally suited for resolving differences in quadrupole coupling constants and accurately determining small asymmetry parameters in materials with several deuterated sites. This is illustrated in Fig. 7, which shows (a) the high field QE spectrum and (b) the zero field NQR spectrum of perdeuterated diethylterephthalate (DETP) [60].

Moreover, zero field spectral lines are sometimes sufficiently narrow to allow deuteron-deuteron dipolar couplings to be resolved. When this is possible, computer simulation of the resulting multiplet patterns permits determination of the relative orientations of the quadrupole and dipole-dipole tensor principal axis systems. In principle, the zero field spectrum can provide equivalent information to that obtained from the analysis of high field, single crystal rotation patterns.

Unfortunately, the technical demands of the zero field NQR technique remain beyond the capabilities of commercially available spectrometers. More fundamentally, this method is limited by the requirement that spins do not relax appreciably during the field switching and sample shuttling times. In practice, the switching and shuttling times are around 20–30 ms and the zero field technique is therefore limited to samples with relaxation times longer than 100–200 ms. This precludes many potentially interesting applications.

An intriguing alternative to the zero field experiment was described by Tycko for $^{13}$C and $^1$H NMR [65]. He showed that, by combining sample rotation with a pulse sequence that accomplishes synchronous rotations in spin space, it is possible to mimic evolution under the full zero field dipolar Hamiltonian in high field (zfhf). The result is a high resolution $^1$H or $^{13}$C spectrum with scaled dipolar splittings but no orientation dependence. This experiment obviates the need to shuttle the sample between different fields and thus does not discriminate against materials with fast relaxation rates. Although considered in an appendix to Ref. [65], the zfhf method has yet to be attempted for deuterons. Two obvious problems are the high duty cycle of the pulses and the need for fast sample spinning. Given that it is not possible to spin the sample fast compared to large quadrupole interactions, analysis of rotational sideband modulation in the zfhf experiment will be necessary. If this proves feasible, the zfhf experiment could become a powerful technique for determining principal values of the quadrupolar tensors for chemically and magnetically inequivalent deuterons.

## *2.8 Two Dimensional Methods*

A generic two dimensional NMR experiment is depicted in Fig. 8. The spins are prepared in some desired state and allowed to evolve for time $t_1$. After the mixing period, signal is recorded in time $t_2$. Double Fourier transformation with respect to both time variables, $t_1$ and $t_2$, yields the two dimensional spectrum.

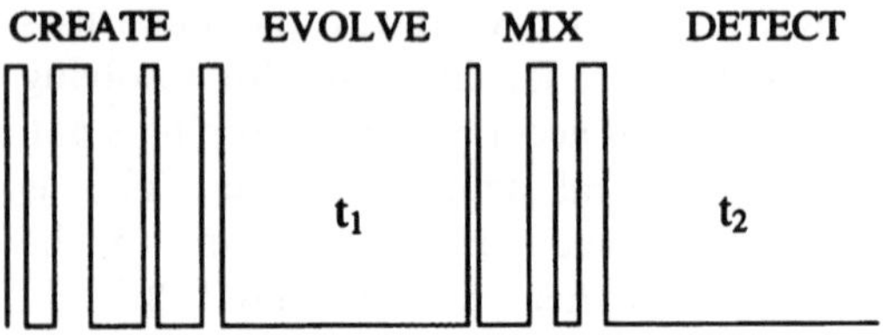


**Fig. 8.** A generic two dimensional pulse sequence. The 2D spectrum is obtained by double Fourier transformation with respect to the evolution, $t_1$, and detection, $t_2$, times

Technically important embellishments for deuteron NMR include the use of an echo pulse to shift the recorded signal out of the dead time after the mixing pulse, and phase cycling to remove undesired coherences. For deuterons 2D methods can be broadly divided into two categories: static and dynamic. Static methods correlate time independent interactions during evolution and observation periods. Potential applications include the measurement of quadrupole coupling parameters for chemically shifted sites, and direct determination of static orientational distribution functions. Dynamic methods make use of the mixing time to allow spin relaxation and molecular motion to occur, and will be discussed in detail in Sect. 4.3.

### *2.8.1 Double Quantum Spectroscopy*

Double quantum (DQ) deuteron NMR provided one of the first demonstrations of the power of 2D methods for detecting unobservable coherences. The importance of double quantum NMR as a structural tool is based on the fact that to first order, quadrupole effects do not shift or broaden the DQ coherence while chemical shifts are magnified by a factor of two. Thus the 2D double quantum method (2D-DQ) competes directly with single quantum MAS. Currently there is insufficient experimental evidence to decide which method is better for obtaining high resolution $^2$H-NMR chemical shift spectra of solids and liquid crystals.

An early experiment by Eckman et al. [38] demonstrated the potential of combining the double quantum and MAS techniques, and 1D versions of the method have recently received attention [66]. By relying on sample spinning to remove the first order quadrupole broadening, the constraints on accuracy and stability of the spinning axis are considerably relaxed. The spinning partially removes deuteron homo- and hetero-nuclear dipolar coupling in the double quantum domain, and detection of the rotational echo train gives greatly enhanced sensitivity. By analogy to the TOPS experiment, an obvious 2D representation would correlate the double quantum signal and the isotropic chemical shift spectrum. This has yet to be exploited as a source of increased resolution.

### *2.8.2 Dynamic Angle Spinning and Double Rotation*

Double quantum and MAS are not the only methods available for correlating anisotropic with isotropic interactions. For half integer quadrupolar nuclei, the closely related techniques of Dynamic Angle Spinning (DAS) and Double Rotation (DOR) have recently proved to be highly successful [67–72]. The aim of these techniques is to use appropriate sample rotations to simultaneously remove both the first and second order quadrupole broadening. In DAS, the sample is spun for time $t_1$ about one axis and the polarization stored parallel to the static field. The spinning axis is rapidly switched to a second orientation and data is collected during time $t_2$. The result is a two dimensional spectrum with isotropic shifts along one axis correlated with second order lineshapes along the other [70]. In the technically more difficult DOR experiment, the sample is spun simultaneously about two axes.

For half integer quadrupolar nuclei the relative merits of DAS and DOR, as methods for obtaining high resolution solid state spectra of the $(1/2 \rightarrow -1/2)$ transition, are currently under investigation in several laboratories [72–75]. These methods have not yet been applied to integer spins, where the spectrum is dominated by the large first order quadrupole interaction. Since it is impossible to spin fast compared to large quadrupole splittings, DAS and DOR spectra of deuterons will generally have a large number of spinning sidebands. This severely complicates the analysis and can degrade the resolution [71].

### *2.8.3 Orientation Distribution Evaluation and Reconstruction*

Direction Exchange with Correlation for Orientation Distribution Evaluation and Reconstruction (DECODER) [76], is similar in concept to DAS. It is essentially a 2D exchange experiment in which the sample is mechanically flipped during the mixing time. This experiment has been demonstrated for both $^2$H [76] and $^{13}$C [77]. For anisotropic interactions with axial symmetry, the intensity distributions in the 2D plane are characteristic of the orientational probability distribution. Figure 9 illustrates the application of DECODER to the crystalline component of a drawn polyethylene sample [76]. The 2D ridge patterns are used to reconstruct the orientational distribution function, which has uniaxial ordering perpendicular to the rotation axis. Since the orientational probability distribution function can be obtained directly from the 2D lineshape [78], DECODER appears to be more informative than methods based on simulation of sideband intensities in slow spinning spectra. Although it has so far been applied primarily to oriented polymers, the experiment has potential for investigations of partially ordered solids and smectic liquid crystals (both thermotropic and lyotropic), where the uniaxial and biaxial order parameters are of interest. It appears that complicated orientational distributions, which are difficult to analyze from the 1D spectrum, can be determined using

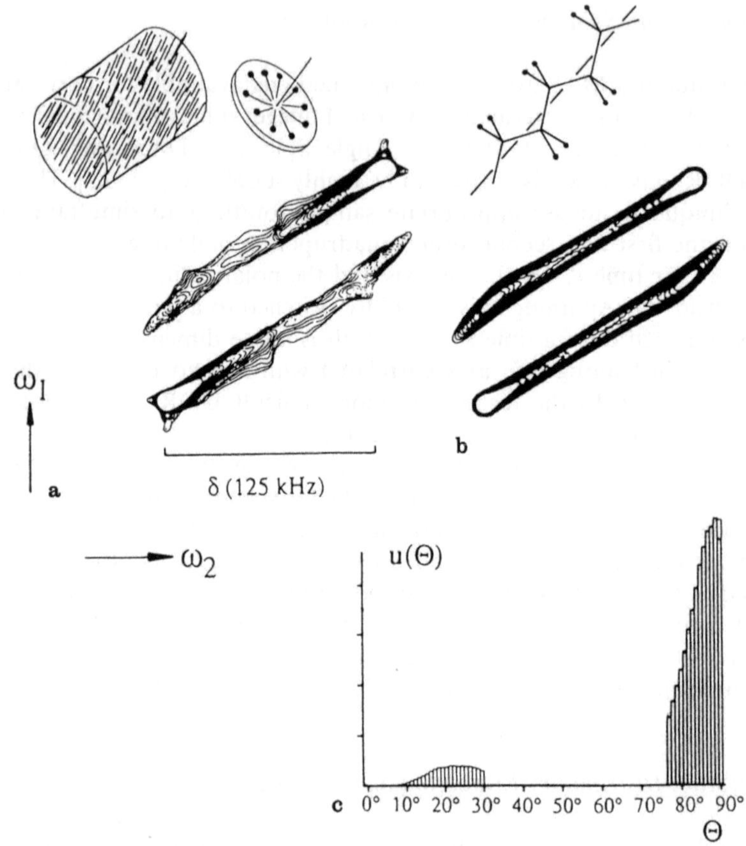


**Fig. 9a–c.** The 2D DECODER NMR spectrum of the crystalline fraction of deuterated polyethylene. The sample reorientation angle was $\beta = 16°$, from $\beta_1 = 38°$ to $\beta_2 = 54°$, and the uniaxial sample has the draw direction perpendicular to the rotation axis. The (**a**) experimental and (**b**) simulated 2D exchange spectra are shown. (**c**) The orientational distribution function $u(\Theta)$ is reconstructed from the experimental spectrum. Reproduced with permission from Schmidt-Rohr et al. [76]

DECODER, with appropriate choice of reorientation angles. However, it must be noted that the zero field, DAS and DECODER methods all suffer from a common problem: because the sample reorientation takes ~50 ms, only samples with long $T_{1Z}$ and slow spin diffusion can currently be investigated.

### 2.8.4 *Variable Angle Correlation Spectroscopy*

Variable Angle Correlation Spectroscopy (VACSY) [75] is an interesting alternative which preserves the anisotropic interactions and provides a 2D correlation with the isotropic chemical shift. In this sample spinning experiment, free induction decays are recorded for time $t_2$ while the spinning axis is stepped through a series of values. Spinning the sample about an axis inclined at $\beta$ degrees to the static field scales all the secular parts of second rank tensor interactions by a factor $P_2(\cos\beta) = (3\cos^2\beta - 1)/2$. This may be regarded as defining evolution under the full (secular) Hamiltonian for a scaled time $t_{aniso} = P_2(\cos\beta)t_2$. While the two dimensional data are collected along orthogonal time axes ($t_{iso}$, $t_{aniso}$), the points are unequally spaced and thus not suitable for Fourier transformation. It has been shown that the data can readily be interpolated onto a rectangular grid, and Fourier transformed to produce a 2D spectrum. This contains the desired correlation of isotropic and anisotropic interactions.

VACSY has provided useful information in $^{13}$C [75] and $^{29}$Si [79] spectroscopy, but its application to deuterons is complicated by the existence of multiple sidebands. These arise because it is impossible to spin the sample rapidly compared with the large quadrupole interaction. Figure 10 shows the VACSY spectrum of diethylterephthalate (DETP) deuterated in the methyl and aromatic positions [80]. The MAS spectrum (shown to the right) is obtained by projection onto the $\nu_2$ axis. Because of the small deuteron chemical shift dispersion, the methyl and aromatic deuterons are barely resolved in the centerband of the MAS spectrum. In principle, the anisotropic quadrupolar interaction for each deuteron could be recovered from corresponding slices in the $\nu_1$ dimension. This is difficult because of the small deuteron chemical shift dispersion, but by

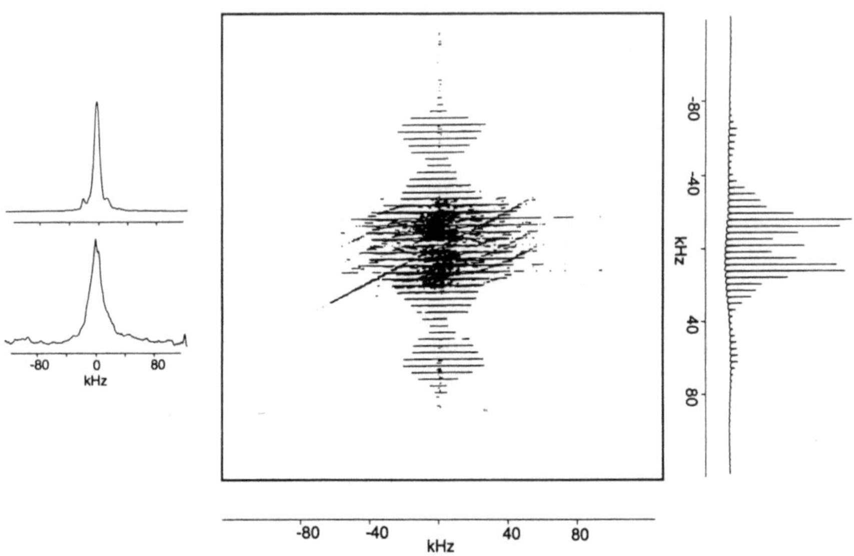


**Fig. 10.** Two dimensional $^2$H-VACSY spectrum of diethylterephalate (DEPT), selectively deuterated on the methyl and phenyl positions. The projection onto the $\nu_2$ axis gives the MAS spectrum, which is shown on the right of the contour plot. Slices (parallel to $\nu_2$) through two specific spinning sidebands are shown on the left. Unfortunately the existence of many spinning sidebands complicates separation of the two anisotropic quadrupole powder patterns [80]

selecting a slice corresponding to points in $\nu_2$ which are outside the methyl powder pattern (bottom trace, left in Fig. 10) it is still possible to isolate the aromatic quadrupolar interaction. However, the existence of multiple sidebands implies that the lineshapes in $\nu_1$ bear little resemblance to ordinary powder patterns [75]. Quantitative determination of quadrupole coupling constants would require detailed simulations of the 2D sideband lineshapes. At present, it appears that the TOPS method [51] is superior to VACSY for correlating deuteron chemical shifts with quadrupole couplings.

In closing this section on static methods, it should be emphasized that most structural information obtained has come from analyzing quadrupole echoes, either to obtain quadrupole coupling tensor parameters or information about orientational distributions. Advanced techniques aimed at suppressing the quadrupole interaction to achieve high resolution are somewhat disappointing, because of the small underlying chemical shift range of deuterons. Nevertheless, when the isotropic shifts can be resolved, the new techniques (especially those using 2D methodology) can permit site specific orientational information to be obtained without recourse to selective isotopic labeling. Some of the techniques, such as DECODER, are capable of yielding the orientational distribution function directly without serious approximations. However, most of the methods (especially one dimensional techniques) have the disadvantage of relying on matching model dependent simulations to experimental spectra.

# 3 Dynamics

## 3.1 *Kinetic Windows*

The power and uniqueness of solid state deuteron NMR studies of molecular motion lie in the wide range of time scales that are accessible for quantitative investigation. Table 1 summarizes the time scales of typical motional processes and the $^2$H-NMR methods that can be used to probe them.

Fast rotations with correlation times in the picosecond range can be investigated using relatively new methods based on the orientation dependence of individual spectral density functions. Processes with correlation times in the

**Table 1.** $^2$H-NMR methods and accessible motional correlation times

| Method | Correlation times |
|---|---|
| 2D Exchange | $10^{-2}$–$10^{4}\,s^{-1}$ |
| Selective Inversion | $10^{-2}$–$10^{5}\,s^{-1}$ |
| Lineshape Analysis | $10^{4}$–$10^{8}\,s^{-1}$ |
| $T_2$ Anisotropy | $10^{-2}$–$10^{4}\,s^{-1}$ |
| $T_{1Z}$ Anisotropy | $10^{8}$–$10^{12}\,s^{-1}$ |
| $T_{1Q}$ Anisotropy | $10^{-1}$–$10^{12}\,s^{-1}$ |

range $10^4$–$10^7$ s$^{-1}$ are usually studied by simulating the effects of motion on quadrupole echo lineshapes. Decay of quadrupolar order, selective inversion, and 2D exchange spectroscopy permit investigation of slow rotational motions. Here the correlation times approach the spin lattice relaxation times, which can range from milliseconds to several hundred seconds. The use of deuteron NMR techniques to study molecular motion has been recently reviewed [81].

## 3.2 Two Dimensional Exchange

Perhaps the most important recent advance in the use of solid state $^2$H-NMR to study dynamics is the introduction of multi dimensional exchange experiments by Spiess and coworkers [82–84]. These techniques provide direct information on the geometry of slow rotational motion without recourse to model dependent simulations. The basic two dimensional pulse sequence is $90_y$-$t_1$-$54.7_j$-$T_m$-$54.7_j$-$\Delta$-$90_x$-$\Delta$-$t_2$. The last $90_x$ pulse and short delays, $\Delta$, are included to produce an echo outside the spectrometer dead time. Two separate sets of experiments (with phase $j = y$ or $x$) must be performed, incrementing the $t_1$ delay at constant mixing time $T_m$. With $j = y$, a two dimensional cosine data set, $\mathrm{cc}(t_1, t_2) = \langle\cos(\omega_1 t_1)\cos(\omega_2 t_2)\rangle$, is obtained while $j = x$ yields a second sine set, $\mathrm{ss}(t_1, t_2) = \langle\sin(\omega_1 t_1)\sin(\omega_2 t_2)\rangle$. Here $\omega_1$ and $\omega_2$ refer to quadrupole splittings during evolution and acquisition times, respectively, and the angular brackets indicate an ensemble average. In principle, both quadrupolar frequencies, $\omega_1$ and $\omega_2$, are time dependent because of molecular motion. However, the experiment is most readily understood when the mixing time is long compared to $t_1$ and $t_2$, so that motion need only be considered during $T_m$. In this case, both $\mathrm{cc}(t_1, t_2)$ and $\mathrm{ss}(t_1, t_2)$ correlate spins whose precession frequency changes from $\omega_1$ during $t_1$ to $\omega_2$ during $t_2$. Double Fourier transformation of $\mathrm{cc}(t_1, t_2)$ and $\mathrm{ss}(t_1, t_2)$ yields spectra whose off diagonal features provide direct information on orientations which are connected by the rotational motion. For single crystal and monodomain liquid crystal samples, this information can be retrieved from either data set, $\mathrm{cc}(t_1, t_2)$ or $\mathrm{ss}(t_1, t_2)$, but for powder samples it is essential to combine the two data sets in order to obtain pure absorption mode spectra in both dimensions.

Wefing et al. [85, 86] have simulated cosine and sine data sets, $\mathrm{cc}(t_1, t_2)$ and $\mathrm{ss}(t_1, t_2)$, for arbitrary values of $t_1, t_2, T_m$ and jump rate for various types of slow molecular motion. They verified the earlier, less general calculations of Schmidt et al. [83] which showed that the 2D method can easily distinguish between discrete rotational jumps and small step diffusive motion. In fact, in favorable cases the characteristic ridge patterns provide enough information to reconstruct the entire one particle, two time conditional probability distribution function for the motion. This is especially important for polymer samples, where the assumption of a unique motional trajectory characterized by a single rate is usually incorrect. Details of how to implement 2D exchange experiments in deuteron spectroscopy have been described by Schmidt et al. [83] and Zhang et al. [87, 88].

The 2D exchange spectrum of bisphenol-A polycarbonate is shown in Fig. 11a. Analysis of the 2D ridge patterns for (b) unstretched and (c) stretched polycarbonate shows that stretching results in a broadening of the distribution of phenyl flipping angles which remain centered about 180° [89].

Diagonal features in deuteron 2D exchange spectra arise from spins whose quadrupole frequencies are constant during the mixing time. Additional contri-

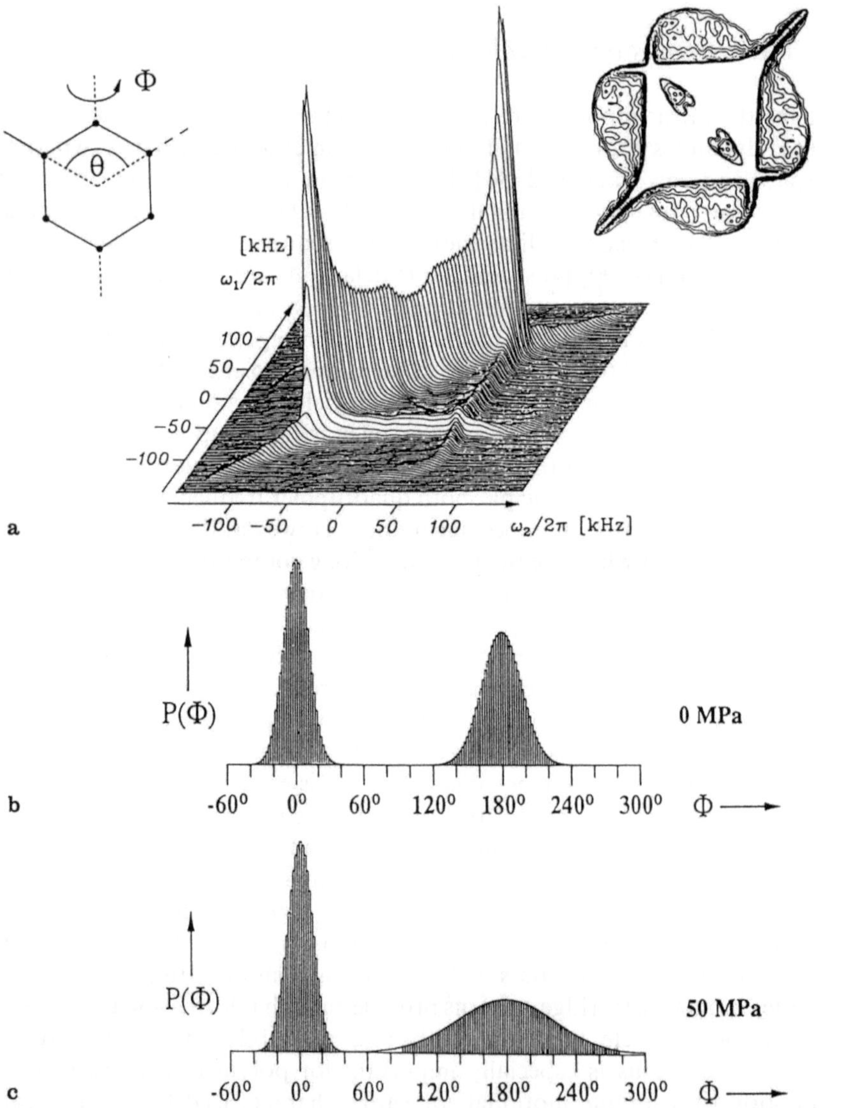


**Fig. 11a.** 2D exchange spectrum for bisphenol-A polycarbonate at 233 K, with a mixing time of 500 ms. The reorientational distribution functions for phenyl ring flips are shown in (**b**) without strain and (**c**) with 50 MPa of strain applied. Reproduced with permission from Hanson et al. [89]

butions arise from spins whose frequencies have changed, perhaps more than once, but are coincidentally the same during $t_1$ and $t_2$ periods. Schmidt-Rohr et al. [2, 90, 91] recently showed how to use reduced three and four dimensional exchange experiments to resolve signals on the basis of this type of kinetic history. 2D methods can be used to investigate motional rates which are too slow to influence the one dimensional, quadrupole echo lineshapes. Moreover, they provide unambiguous, model free information about the trajectory of rotational motion. Their only disadvantage is that accurate determination of jump rates requires analysis of cross peak intensities as a function of mixing time. The data acquisition time needed to obtain complete 2D data sets for a range of mixing times can be prohibitively long.

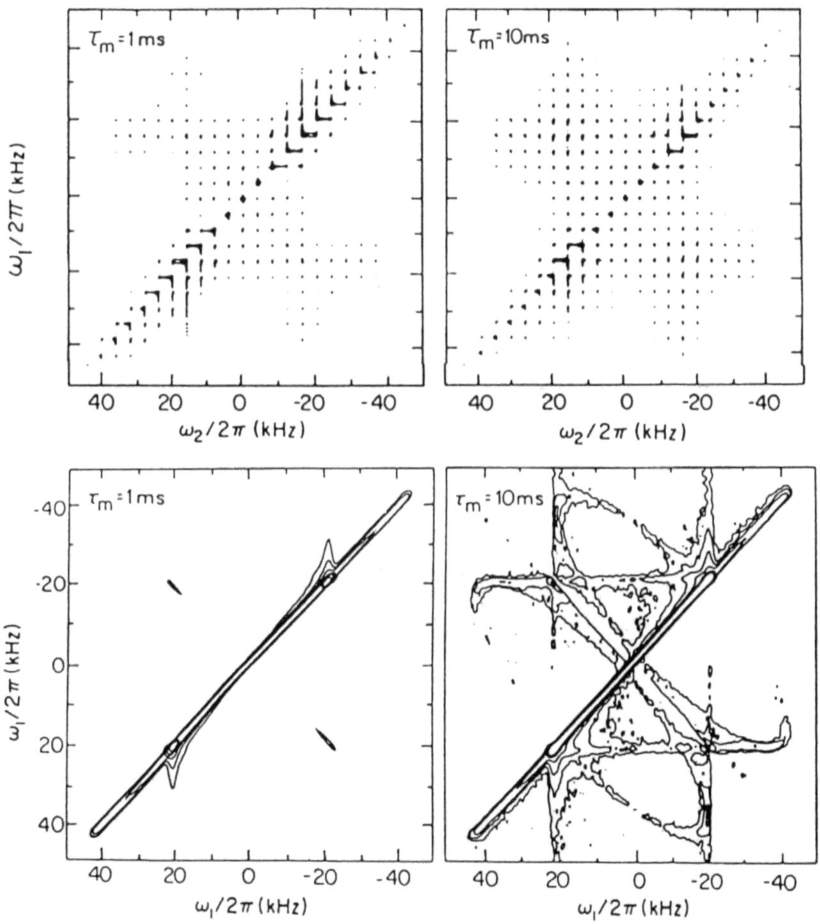


**Fig. 12.** Two-dimensional exchange spectra of dimethylsulfone (DMS) at 30 °C, for mixing times $\tau_m = 1$ and 10 ms. The *top two spectra* show the results of the MAS rotor synchronized experiment ($\nu_r = 4.06$ kHz), while the *bottom two* are for the static sample. Adapted with permission from Reichert et al. [46]

Boeffel et al. [92, 93] described modifications of the 2D sequence which make it more suitable for studies of exchange in oriented liquid crystalline media. It was shown that extra phase cycling procedures are needed to suppress unwanted coherence from transverse magnetization which would otherwise persist during the mixing period, and ultimately yield spurious cross peaks. This experiment is discussed in more detail in the thermotropic liquid crystal Sect. 4.4.1.

Recently Reichert et al. have extended the deuteron 2D exchange method to include magic angle spinning [46]. The rotational echo train is collected in the detection period, and Fourier transformation in this dimension gives a series of sidebands whose intensities trace out the deuteron powder pattern. The mixing time is synchronized with the rotor period and this results in the complete absence of cross peaks in the 2D spectrum unless chemical exchange is occurring [94]. Theoretical expressions were derived for cross peak intensities as a function of the kinetic parameters, and the method was demonstrated for dimethylmalonic acid-$d_6$, dimethylsulfone-$d_6$ (DMS) and a thiourea/cyclohexane-$d_{12}$ inclusion compound [46]. The authors conclude that this techniques main advantage is increased intensity of the cross peaks which, unfortunately, is gained at the expense of losing the characteristic ridge patterns of the static 2D exchange spectra. This feature is illustrated for DMS at 30 °C in Fig. 12. The 2D exchange spectra are compared with (top) and without (bottom) magic angle spinning, for mixing times of 1 and 10 ms.

The MAS 2D exchange technique has the potential to be very important for investigating dynamic processes in solids.

## 3.3 Selective Inversion

Since the earliest days of condensed matter NMR selective excitation (hole burning) has been used to distinguish between homogeneously and inhomogeneously broadened lines [95]. Duncan et al. [96] and Kuhns and Conradi [97] demonstrated the utility of the method for studying slow motion effects on $^{13}C$ lineshapes broadened by chemical shift anisotropy. Applications to deuteron NMR have been reported by Vold, Hoatson and Brown [98], Lin and Vold [99], and Williams and McDermott [100].

Selective inversion (SI) of a narrow band of frequencies can be achieved either by means of a long, weak inversion pulse or by a DANTE sequence of hard pulses with small flip angle [101]. If molecular motion occurs in a time short compared with $T_{1Z}$, the inverted spins can jump to orientations (quadrupole frequencies) outside the excited region. This results in the hole filling in at rate faster than the nonselective $T_{1Z}$ and the difference can be used to determine the jump rate, $k$. Recovery of the hole is accomplished by exchange of inverted magnetization with frequencies unaffected by the selective pulse, and images of the hole appear at frequencies related by the jump angle [98]. Thus in principle, selective inversion provides the same information as the 2D exchange experiment and allows both the trajectory and the rate of slow molecular motion to be determined.

The sequence used for soft pulse selective inversion of deuteron powder patterns is shown in Fig. 13. In this experiment it is necessary to distinguish between on resonance and off resonance excitation. Figure 13a shows the on resonance case ($\Delta = 0$), where the DMS powder pattern has a hole burnt in the center ($\beta = 54.7°$). Since the quadrupole splitting is zero it is tempting, but

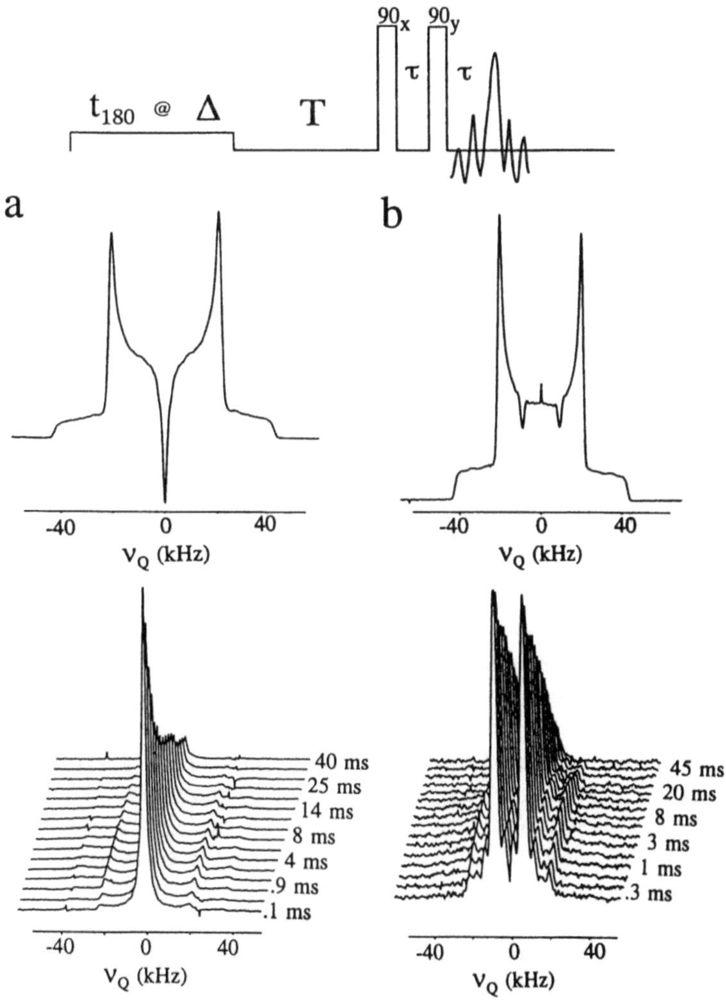


**Fig. 13a, b.** The pulse sequence for the deuteron selective inversion experiment, with quadrupole echo detection. The dimethylsulfone (DMS) powder patterns, with selectively excited range of frequencies, are shown for (**a**) on resonance ($\Delta = 0$) and (**b**) off resonance ($\Delta = 10$ kHz) excitation. On resonance, the low power 180 pulse length is $t_{180} = 400\,\mu s$ and at high power the 90 pulse length is $t_{90} = 1.7\,\mu s$. The *bottom traces* are difference spectra shown as a function of the relaxation delay, $T$, whose values are indicated. The decay time is $T_{1Z}$(sel), which can be compared to the nonselective $T_{1Z}$ to give the jump rate incorrect, to ignore $H_Q$ when calculating the inversion. This procedure would fail to account for the excitation of double quantum coherence by a weak, resonant pulse [40]. Nevertheless, it is still correct to calculate the inversion pulse width $t_{180}$ using the simple relation $\omega_1 t_{180} = \pi$. Off resonance ($\Delta \neq 0$), the fictitious spin-1/2 formalism [102] shows that the modified relation, $\sqrt{2}\omega_1 t_w = \pi$, must be used to determine the correct 180° pulse width. Moreover, the asymmetrical excitation, created by selective inversion on one side of the spectrum, is mixed by nonlinear effects of the strong pulses used in the quadrupole echo detection sequence. Thus the powder and difference spectra shown in Fig. 13b are symmetric. In the bottom traces, both the on and off resonance difference spectra are plotted as a function of relaxation delay. Analysis of the recovery gives the selective $T_{1Z}$(sel) from which the jump rate can be determined. Also apparent in the on resonance difference spectra, Fig. 13a, is a gradual increase in intensity near the perpendicular edges of the powder pattern, followed by a slow decay. This is expected because the 106° jump angle [83, 103] transfers polarization from the center of the spectrum to this orientation.

The SI experiment would be improved by the introduction of shaped pulses. It is anticipated that these will give better selectivity and inversion efficiency. Because of the mixing effects noted above, the most effective shaped pulses are likely to be those that create symmetric selective excitation of the spectrum. The DANTE sequence [101] is an example of this type of excitation. Warren [104] discusses general procedures for designing shaped pulses, but these promising ideas have not yet been widely exploited in solid state deuteron NMR studies.

Theimer and Bodenhausen [105] have compared selective inversion with 2D methods as a source of kinetic information from $^{13}C$ powder patterns. They reached the rather negative conclusion that one dimensional selective inversion is less useful than the 2D approach. It should be appreciated that this conclusion is probably irrelevant for deuterons, primarily because the interactions strengths, relaxation times, and associated kinetic windows are very different. In particular, quadrupole broadened $^2H$ powder patterns are usually 10–100 times wider than $^{13}C$ chemical shift anisotropies, and the underlying homogeneous linewidth is on the order of 1 kHz. This makes it is possible to achieve complete inversion over the homogeneous linewidth with a 500 μs pulse. $T_{1Z}$ and $T_{1Q}$ are usually long enough to ignore during the inversion, and exchange rates as small as 10–50 $s^{-1}$ can be measured accurately [106]. It appears that for deuterons selective inversion offers a quick, quantitative way to measure the motional rate, while 2D exchange experiments yield more direct information on motional trajectories.

### 3.4 *Lineshape Analysis*

Quadrupole echo lineshapes, recorded as a function of temperature and pulse spacing, continue to provide remarkably detailed and quantitative information

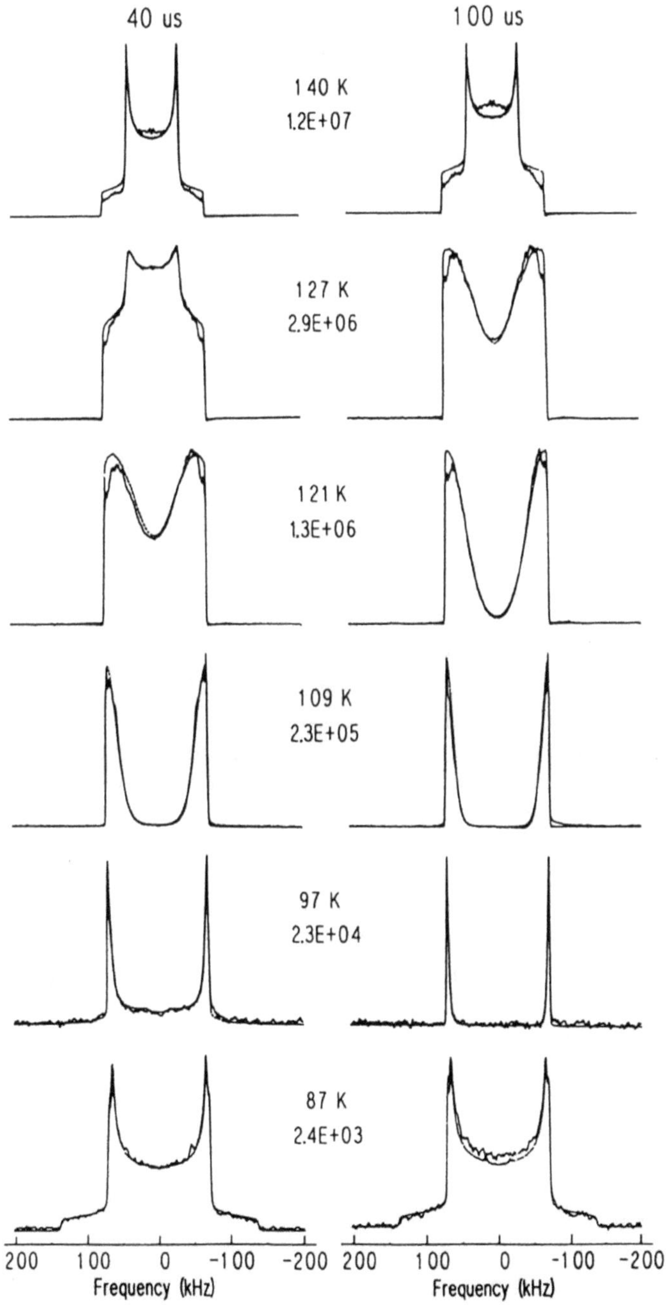


**Fig. 14.** The temperature dependence of the quadrupolar echo spectra of solid benzene, for two pulse spacings $\tau = 40\,\mu s$ (*left*) and $\tau = 100\,\mu s$ (*right*). Experimental lineshapes (*solid lines*) are superimposed on simulations (*dashed lines*) to determine the $C_6$ jump rate. Reproduced with permission from Ok et al. [108]

about simple rotational motions in solids [107]. The slowest rates accessible by this technique ($\sim 10^4\,s^{-1}$) are limited by the underlying homogeneous linewidth of the inhomogeneously broadened powder pattern. The fastest ($\sim 10^8\,s^{-1}$) are limited by the requirement that the rate be comparable to the difference in quadrupole splitting between orientations connected by the motion. A particularly attractive feature of QE lineshape analysis is that the transverse relaxation time $T_2$ is strongly orientation dependent. The dependence of the spectrum on the pulse spacing is highly characteristic of the motion and can be used to discriminate between different models. An example of using lineshapes to derive kinetic information is given for solid benzene [108] in Fig. 14. The experimental quadrupolar echo lineshapes were compared to simulations to deduce the $C_6$ jump rate, and the temperature dependence was used to determine the activation energy of motion.

A considerable body of recent work has been devoted to deriving information about complex motional processes by means of simulations. In polymer systems simulations must incorporate a broad distribution of rate constants, and perhaps of quadrupole coupling constants and site populations. In their extensive experiments and simulations of dynamics in Nylon 66 and related materials, Hirschinger et al. [109–111] demonstrated that great care is needed in interpreting spectra of systems with complicated motional processes. In particular, they found for rapid motion that the lineshapes with temperature dependent, site specific populations, might in some circumstances be mistaken for intermediate motion with equal populations and a temperature dependent distribution of rates. This work highlights one of the weak aspects of lineshape analysis: the need for model dependent simulations. Even when a particular models' parameters have been adjusted to reproduce the spectra, it is always possible that another, very different model may fit the data equally well. Hirschinger and English [109] recommend that in order to avoid misinterpretations, the minimum precaution is to require that any model fit not only the lineshape and it's pulse spacing dependence, but also the anisotropy in spin lattice relaxation time $T_{1Z}$. As will be discussed in Sects. 3.6 and 3.7, it is now possible to measuring both $T_{1Z}$ and $T_{1Q}$ in powders. This allows the individual spectral densities $J_1(\omega_0)$ and $J_2(2\omega_0)$ to be determined for all points on the lineshape [112]. Requiring the orientation dependence of the individual spectral densities to be reproduced places additional stringent constraints on acceptable descriptions of molecular motion.

## 3.5 $T_2$ Anisotropy

In the absence of slow motion, the transverse relaxation time is given by [81]

$$\frac{1}{T_{2e}} = \frac{3\pi^2}{4}(e^2 q_{ZZ} Q/h)^2\,(3J_0(0) + 3J_1(\omega_0) + 4J_2(2\omega_0)) \tag{16}$$

where $T_{2e}$ denotes the transverse relaxation time measured from the pulse spacing dependence of the quadrupole echo sequence. When slow motion is present, $T_{2e}$ can be strongly anisotropic across the powder pattern and this leads to characteristic distortions of the quadrupole echo at large pulse spacing. Kothe and coworkers [113–115] have emphasized the utility of two dimensional techniques to determine the anisotropy of the transverse relaxation rate, $T_{2e}$,

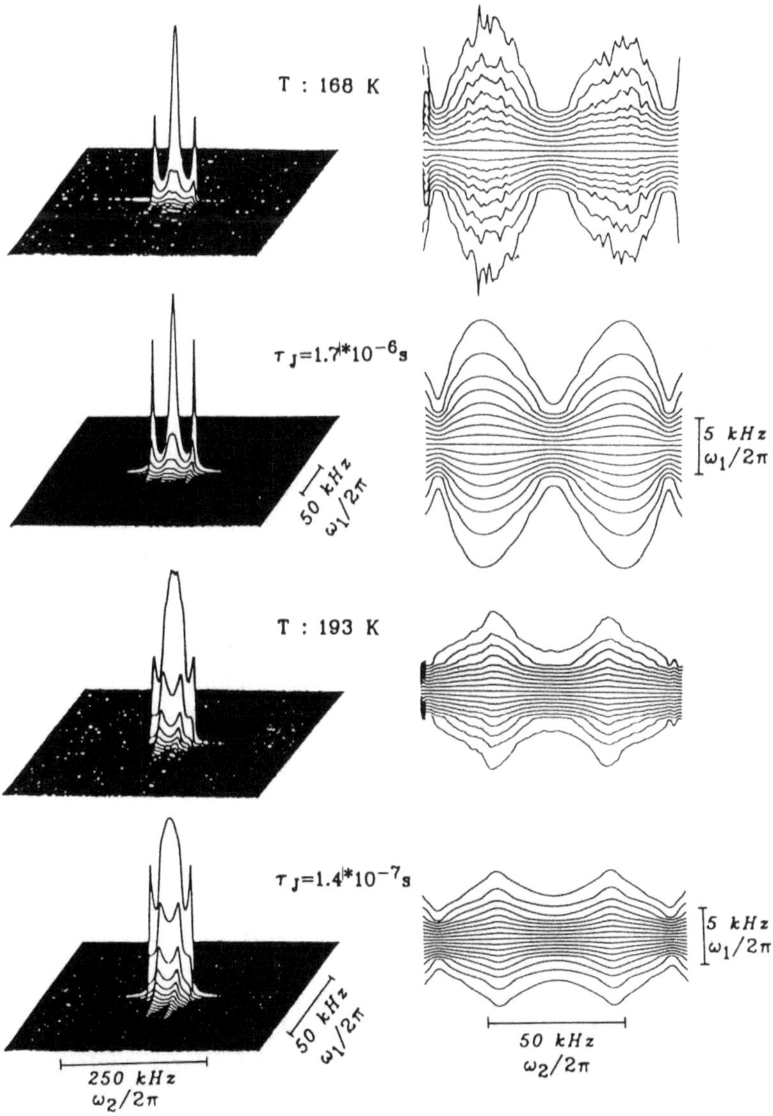


**Fig. 15.** 2D NMR relaxation spectra (quadrupole echo sequences) of methyl deuterated L-alanine-$d_3$, shown at temperatures 168 and 193 K. Stack plots are shown on the *left* and normalized contour plots on the *right*. The simulations were obtained using a three site jump model, with the correlation times shown, $\tau_J$. Reproduced with permission from Schleicher et al. [115]

in this motional regime. In this approach, the evolution time $t_1$ is the time from the first pulse to the top of the quadrupole echo, where the start of data acquisition initiates $t_2$. Contour maps of the doubly Fourier transformed data set yield convenient and characteristic displays of the anisotropy of $T_{2e}$. This method is illustrated in Fig. 15, which shows the experimental and simulated 2D relaxation spectra for the methyl group in L-alanine-$d_3$, at two temperatures [115]. The appearance of the 2D contour plots is highly characteristic of the rate and type of slow molecular motion and allows kinetic parameters to be determined.

The multiple pulse quadrupolar echo sequence (MQE) [116], $90_x$-$\tau$–$90_y$-($2\tau$–$90_y$-)$_n$-, can also be used to measure the transverse relaxation time, $T_2$. In this experiment is important to realize that $T_2$ depends on the pulse spacing, $2\tau$. Determining this dependence is most informative because analytical expressions exist for $T_2$ for both limiting cases; $\tau \to 0$ and $\tau \to \infty$ (Eq. 16). The limiting relaxation rates are different linear combinations of the individual spectral densities $J_0(0)$, $J_1(\omega_0)$ and $J_2(2\omega_0)$. At small pulse spacing, there is insufficient time for differential precession of the two one-quantum coherences and the spins are effectively spin locked. Under these circumstances transverse cross relaxation is a secular process and shortens $T_2$. At large pulse spacing ($\tau > 1/\nu_Q$), transverse cross relaxation is suppressed. Ahmad et al. [116], and Mueller et al. [117, 118] have discussed the use of the MQE sequence in ordered fluids to study chemical exchange. In practice, it is difficult to make the pulse spacing small enough to suppress cross relaxation without overheating the sample. For large $\tau$ some caution must be exercised since unresolved dipolar coupling can contribute to the apparent $T_2$ [81].

With appropriate precautions, the multipulse quadrupole echo method is worth considering because it is a very efficient way to measure $T_2$s of individual lines in a multiline spectrum. A 2D version of the technique is illustrated in Fig. 16, for fluorene-$d_{10}$ dissolved in a nematic liquid crystal 5CB.

Here the 2D data set was obtained by collecting the last half echo as a function of the number of $90°_y$ pulses in the sequence. Double Fourier transformation with respect to the acquisition time $t_2$ and the pulse train duration $t_1 = 2n\tau$ gives the 2D spectrum. For fluorene the pulse spacing was $2\tau = 300\,\mu s$ and every fifth echo was recorded ($n = 1, 6, 11, \ldots$). The ordinary QE spectrum is obtained as a slice along $\nu_1 = 0$, and each peak has a Lorentzian lineshape in the $\nu_1$ dimension. The individual $T_2$ values are obtained from the linewidths at half height in the $\nu_1$ domain.

Measurements of $T_2$ are important because they provide the only means of measuring the zero frequency spectral density, $J_0(0)$. This is particularly important in liquid crystalline materials, which may have large contributions from collective processes, such as order director fluctuations (ODF) [119] in nematics and undulation modes in smectics [120]. Like $J_1(\omega_0)$ and $J_2(2\omega_0)$, $J_0(0)$ is inherently orientation dependent in solids and anisotropic fluids. However, aside from measurements on small solutes in nematic liquid crystal

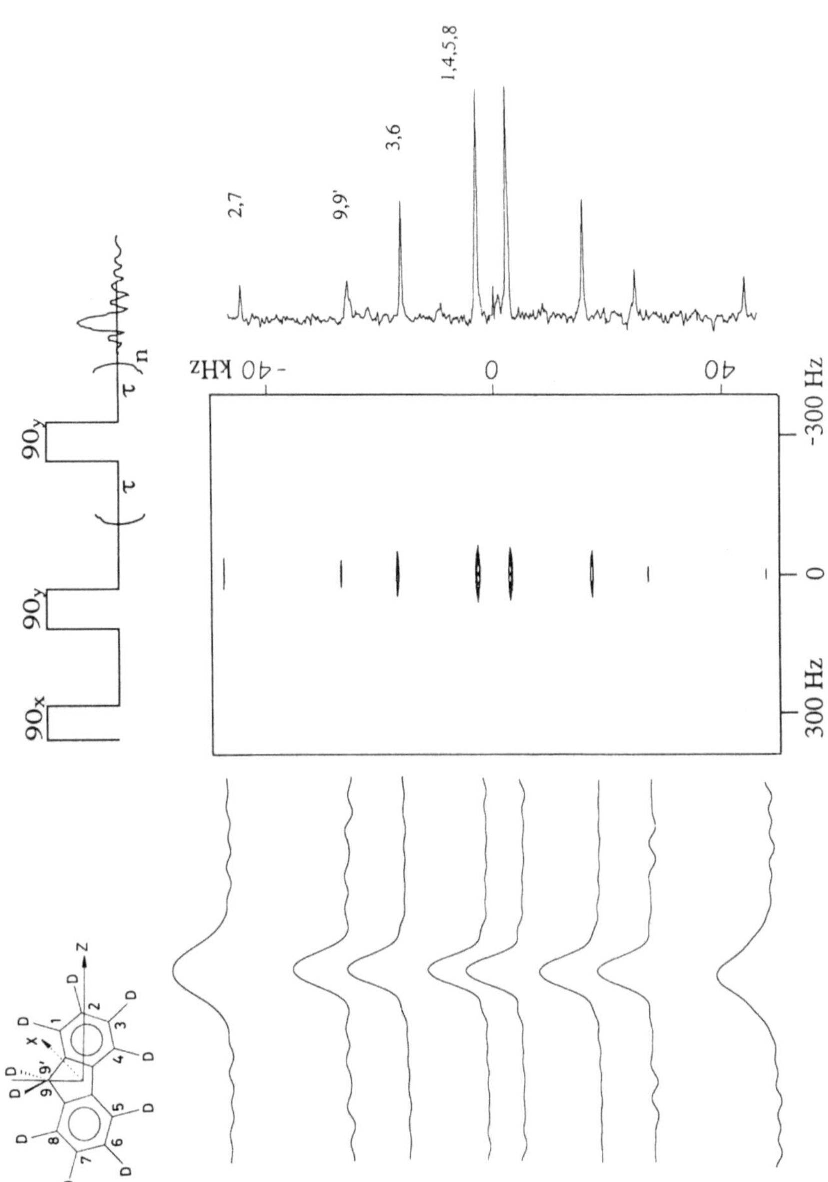


**Fig. 16.** Two dimensional method for determining $T_2$ using the multiple pulse quadrupolar echo sequence. The sample is perdeuterated fluorene in the nematic phase of 5CB. The data were collected with a pulse spacing of $2\tau = 300\ \mu s$ and every fifth echo was recorded ($n = 1, 6, 11 \cdots 76$). The normal quadrupolar echo spectrum is obtained from the $\nu_1 = 0$ slice through the data matrix and is shown *on the right* of the contour plot. $T_2$ values of individual lines are obtained from the corresponding $\nu_2$ slices, shown *on the left*, by measuring the full width at half height of the Lorentzian lineshapes

solvents [121–124] and studies by Kothe and coworkers of liquid crystalline polymers [114, 120, 125, 126], reports of $J_0$ (0) determined from $T_2$ are very scarce.

## 3.6 $T_{1Z}$ Anisotropy

For fast motion, when motional correlation times are on the order of the inverse of the Larmor frequency ($10^8$-$10^{12}\,s^{-1}$) the Zeeman spin lattice relaxation time is given by the equation

$$\frac{1}{T_{1Z}} = \frac{3\pi^2}{2}(e^2 q_{ZZ} Q/h)^2 (J_1(\omega_0) + 4J_2(2\omega_0)) \tag{17}$$

$J_1(\omega_0)$ is the spectral density function corresponding to single quantum spin flips, and $J_2(2\omega_0)$ to two-quantum spin flips. Both functions are orientation dependent in solids, and since the powder pattern is an inhomogeneous superposition of spectra from randomly oriented crystallites, the $T_{1Z}$ varies across the lineshape. Figure 17 shows partially relaxed spectra of HMB, following an inversion recovery experiment with QE detection. By fitting the recovery curves $T_{1Z}$ can be determined as a function of frequency, as shown in the graph [112].

In early measurements of $T_{1Z}$ anisotropy [127], it was common practice to focus attention on only the horns ($\beta = 90$) and the shoulders ($\beta = 0$) of the powder pattem and define an anisotropy factor in terms of measurements at only these two points. While expedient, this procedure is not as informative as measuring the full orientation dependence, and in some cases it can lead to unsuspected systematic error. In particular, if the initial 180° pulse in an inversion recovery sequence does not invert the powder pattern perfectly, then spin diffusion and other slow motional processes can transport partially inverted magnetization from one part of the lineshape to another, and this contributes to the apparent $T_{1Z}$ [128]. For this reason, it is particularly important to use composite inversion pulses to measure $T_{1Z}$ anisotropy of wide powder patterns.

Torchia and Szabo [129] developed a formalism based on Redfield relaxation theory which has proven extremely successful for interpreting relaxation due to fast motion. This includes limited amplitude librational motion in addition to rotational jumps and small step rotational diffusion. The stochastic Liouville equation has been used by Wittebort et al. [130] and Kothe and coworkers [115] to extend these calculations beyond the Redfield limit. General procedures have been developed for calculating spin lattice relaxation due to a combination of discrete jumps and small step diffusion, and have been reviewed by Vold and Vold [81].

## 3.7 $T_{1Q}$ Anisotropy

For an isolated deuteron quadrupole order can be created by means of the Jeener-Broekaert pulse sequence $90_x$-$t$-$45_y$ [131]. The density matrix for a

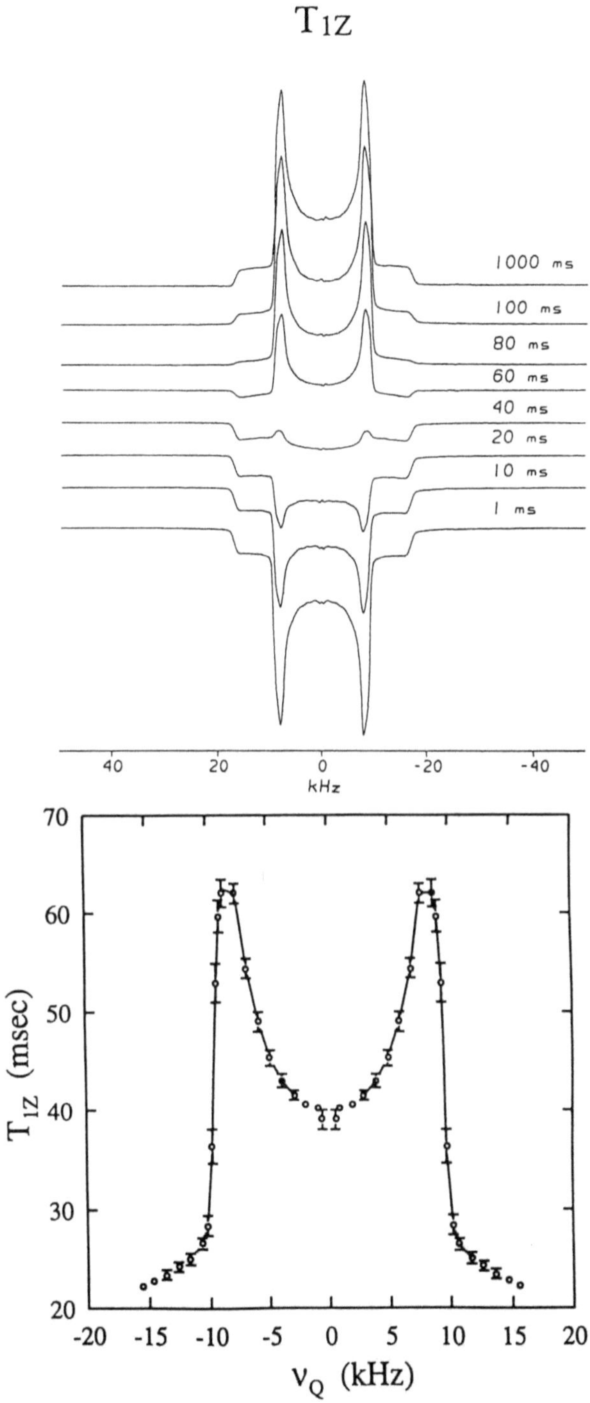


**Fig. 17.** Partially relaxed spectra of hexamethylbenzene (HMB) obtained from the inversion recovery sequence with quadrupolar echo detection (IRQE). The relaxation delay, $T$, is shown *on the right* of the traces. Analysis of the recovery curves from many points on the lineshape gives the anisotropy profile of the relaxation time, $T_{1Z}$, shown in the graph

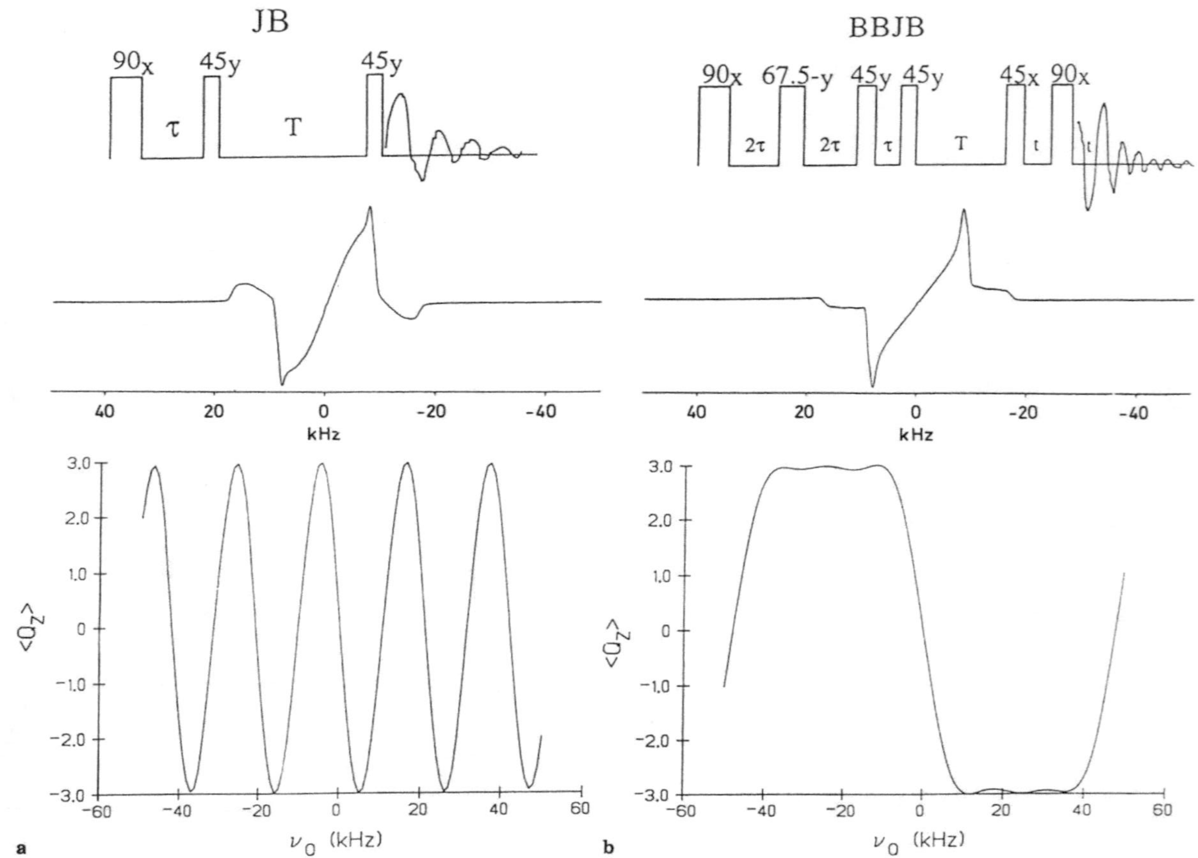


**Fig. 18a, b.** A comparison of the (**a**) conventional (JB) and (**b**) broadband (BBJB) Jeener Broekaert sequences [28]. Theoretical quadrupolar order excitation profiles, including finite pulse width effects, are shown at the *bottom* of the figure. These are calculated for excitation times of (**a**) $\tau = 46\,\mu s$ and (**b**) $5\tau = 46\,\mu s$. The non-uniform excitation of the JB sequence yields a highly distorted powder pattern quadrupole ordered state is diagonal, with elements (1,-2,1) corresponding to components of the quadrupolar doublet pointing parallel and antiparallel to the field. For this reason, quadrupole order is often called spin alignment [132]. The conventional Jeener-Broekaert sequence is inherently frequency selective. In single crystals and monodomain liquid crystals, where the spectra consist of a few well resolved quadrupolar doublets, it is usually easy to maximize the creation of quadrupole order for any given doublet by adjusting the excitation time $\tau$ such that $\nu_Q\tau = (2n+1)/4$. In these systems, decay of quadrupole order during a variable relaxation delay can then be monitored by applying a third $45_y$ pulse, Fourier transforming the resulting antiphase signal, and measuring the difference magnetization.

For powder samples, the continuous distribution of quadrupolar splittings implies that it is impossible to create uniform excitation. It can be shown that at time $\tau$ after the monitoring pulse, all the antiphase doublets will have refocussed along a single axis in the rotating frame, leading to the Jeener-Broekaert or spin alignment echo [132]. Fourier transformation starting from the top of the echo yields a distorted spectrum, in which the powder pattern is convoluted with a $\sin^2(\nu_Q\tau)$ function. Thus, $T_{1Q}$ can only be obtained from points on the spectrum for which quadrupolar order has been excited and the procedure suffers from low accuracy.

For solids a better procedure is to use a broadband composite excitation sequence [27, 28], including an echo pulse to move the resulting signal beyond the receiver dead time (BBJB). This has been shown to create quadrupolar order over a full width powder pattern [28] and to give more accurate and reliable results [133]. Figure 18 shows a comparison of the (a) conventional (JB) and (b) broadband (BBJB) sequences. The spectra are from HMB with identical excitation delays; (a) $\tau = 46\,\mu s$ and (b) $5\tau = 46\,\mu s$. Directly below the spectra are simulations of the excitation profiles for quadrupolar order, including the finite pulse widths.

In the Redfield regime, $T_{1Q}$ is a direct measure of the single spectral density $J_1(\omega_0)$ according to the relation

$$\frac{1}{T_{1Q}} = \frac{9\pi^2}{2}(e^2 q_{ZZ} Q/h)^2 J_1(\omega_0) \tag{18}$$

It is important to recognize that this equation is not valid for slow motion, $\tau_c \sim (1/\nu_Q)$. In this regime additional contributions to $T_{1Q}$ occur and the amplitude of the spin alignment echo is a strong function of the motional rate. This behavior is the physical basis of the 2D exchange experiment described in Sect. 3.2. Only for fast motion the relaxation rates can be used to determine spectral densities of motion.

Partially relaxed spectra of HMB, obtained using the BBJB sequence, are shown in Fig. 19. A two parameter non linear least squares fit of the recovery can be used to determine $T_{1Q}$ for all points on the powder pattern. This anisotropy is shown in the lower part of the figure.

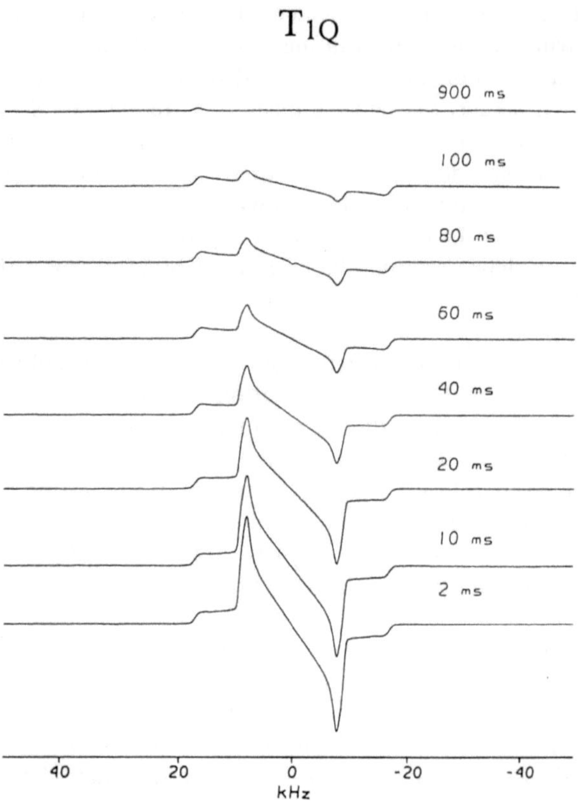
T1Q
900 ms
100 ms
80 ms
60 ms
40 ms
20 ms
10 ms
2 ms
40
20
0
-20
-40
kHz

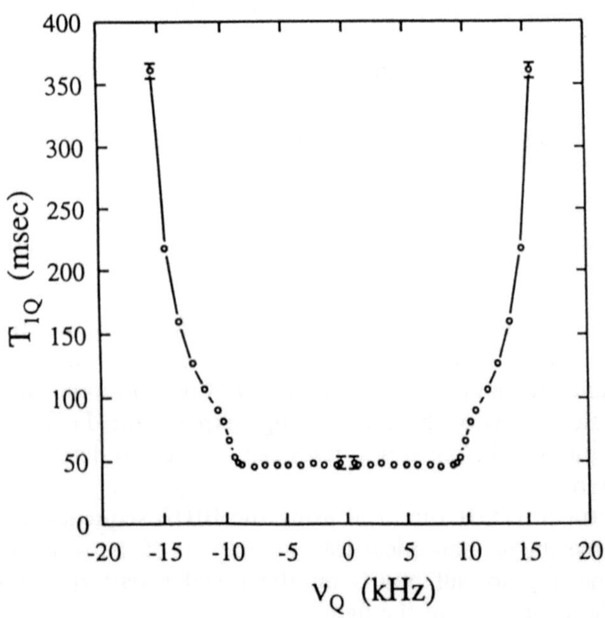
400
350
300
250
200
150
100
50
0
T1Q (msec)
-20
-15
-10
-5
0
5
10
15
20
νQ (kHz)

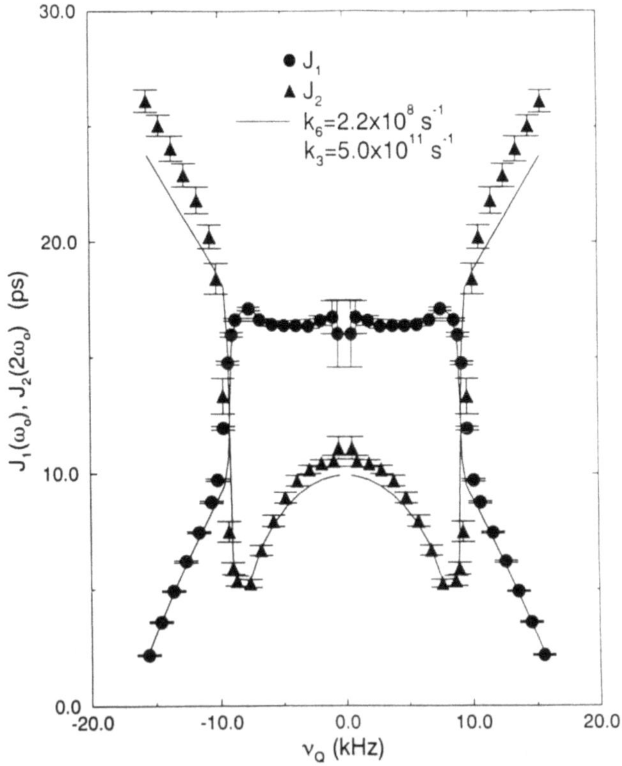


**Fig. 20.** Determination of both relaxation times, $T_{1Z}$ and $T_{1Q}$, allows the individual spectral densities of motion, $J_1(\omega_0)$ and $J_2(2\omega_0)$, to be derived. Experimental values are shown for HMB, these were obtained from the data shown in Figs. 17 and 19. The *solid lines* are the best fit to simultaneous three-fold methyl jumps (at rate $k_3$) and six site jumps around the ring normal (at rate $k_6$). Adapted with permission from Hoatson et al. [112]

By simultaneously measuring $T_{1Z}$ and $T_{1Q}$, it is possible to obtain accurate, model independent measurements of both spectral densities. Such measurements contain considerable additional information on molecular motion in solids and liquid crystals. Measurements of individual spectral density anisotropies across a powder lineshape are extremely sensitive to the details of molecular motion. Recently, they have been used to discriminate subtle differences in motional models for polycrystalline hexamethylbenzene [112]. These results, shown in Fig. 20, were used to determine both the rate of methyl spinning, $k_3$, and the slower rate of rotation about an axis normal to the ring plane, $k_6$.

---

**Fig. 19.** Partially relaxed spectra of hexamethylbenzene (HMB) obtained from the broadband Jeener Broekaert sequence with echo detection. The relaxation delay, $T$, is shown *on the right* of the traces. Analysis of the recovery curves from many points on the lineshape gives the anisotropy profile of the relaxation time, $T_{1Q}$, shown in the graph

# 4 Application to Specific Systems

## 4.1 *Molecular Crystals*

In this section representative $^2$H-NMR studies of molecular crystals are reviewed. These serve to establish typical values of quadrupole coupling parameters for important functional groups, their dependence on molecular structure, and the effects of different kinds of molecular motion. In addition, these examples illustrate the scope and limitation of theoretical descriptions of order and motion commonly used to interpret spectra and relaxation data.

### *4.1.1 Methyl Groups*

Quadrupole coupling constants for methyl groups (Table 2) typically lie in the range 45–55 kHz, and the asymmetry parameter is often zero within experimental error ($\pm 0.01$). The small QCC is a result of fast rotational motion around the pseudo-threefold spinning axis; for tetrahedral geometry the static QCC value is reduced by a factor of $-1/3$. While there is considerable variation in the activation energy for methyl rotation, it is inevitably fast at room temperature.

In molecules such as hexamethylbenzene-$d_{18}$ (HMB) [134, 135], dimethylmalonic acid-$d_6$ [136], and thymine-$d_3$ [137], thermally activated methyl rotation can be frozen out by cooling the sample below $-100\,°C$. Less hindered methyl rotation is observed in acetone-$d_6$ and copper acetate-$d_3$ monohydrate [138]. In general, low temperature methyl spectra do **not** necessarily provide

**Table 2.** Methyl deuteron quadrupole coupling constants

| Substance | QCC | $\eta$ | Temperature | Method[1] | Reference |
|---|---|---|---|---|---|
| acetic acid | 56.3 | 0.088 | 77 K | CW | 356 |
| acetone | 55.4 | 0.144 | 77 K | CW | 356 |
| acetonitrile | 53.2 | 0.0 | 77 K | CW | 356 |
| dimethylsulfone | 59.0 | 0.0 | 77 K | CW | 357 |
| Hexamethylbenzene | 22.7[2] | 0.085 | 298 K | QE | 112 |
| thymine | 51.9 | 0.075 | ambient | QE | 137 |
| thymidine | 50.4 | 0.09 | ambient | QE | 142 |
| BOTI[3] | 51.9 | 0.13 | ambient | QE | 143 |
| POTI[4] | 51.6 | 0.14 | ambient | QE | 143 |
| acetanilide | 54.0 | 0.10 | 284 K | QE | 144 |
| alanine | 55.0 | 0.0 | 298 K | QE | 146 |
| dimethylterephthalate | 50.7 | 0.0 | ambient | ZF | 358 |

[1] CW = continuous wave (field or frequency sweep); QE = quadrupole echo powder lineshape; ZF = zero field experiment
[2] Multiply this number by two to account for rapid rotation about the ring normal
[3] BOTI = (1,1,1,3,5,7-$d_6$)*N*-butyl-2,4,6 octatrienylideneimine
[4] POTI = (1,1,1,3,5,7-$d_6$)*N*-pentyl-2,4,6 octatrienylideneimine

an unambiguous determination of static quadrupole coupling parameters. Instead, they exhibit fine structure and broadening due to quantum mechanical tunneling [138–140]. Relations between the diamagnetic susceptibility and deuteron quadrupole coupling tensors of methyl halides have been studied by van Zijl and Bothner-By [141]. For the halomethanes, the data indicate good agreement between QCCs found from gas phase microwave studies, low temperature solid state NMR and NQR experiments, and certain liquid crystal mixtures.

Rapid rotation about an axis of threefold or higher symmetry necessarily yields an average axially symmetric electric field gradient tensor. Thus observations of $\eta > 0.05$ for HMB [134] and thymine [137] are perplexing, at first glance. In HMB, the large asymmetry parameter is thought to be associated with distorted motion [22, 112], while in thymine and [137] and thymidine [142] it has been ascribed to an intrinsic asymmetry of the electric field gradient tensor. Wann and Harbison recently pointed out that small, non zero values of $\eta$ are to be expected for methyl groups attached to any framework with less than threefold symmetry [143].

It is tempting to postulate a reference value for the static quadrupole coupling constant of a rigid methyl group, in order to interpret observed spectra in terms of different types of fast motion. For example, if the static QCC is assumed to be 165 kHz, then fast spinning (with tetrahedral methyl geometry) reduces the apparent QCC to 55 kHz. Observation of significantly smaller values can then be ascribed to additional fast motion, such as restricted wobbling of the methyl spinning axis. Unfortunately, quantitative estimates based on such restricted motion are highly sensitive to the choice of the static reference value. This is especially true when the angular extent of restricted wobbling is small ($\leqq$ 10–20°). Under these circumstances a better procedure is to measure the temperature dependence of the QCC, which is expected to decrease as the wobble angle increases at higher temperatures.

Spin lattice relaxation of deuterated methyl groups tends to be dominated by threefold rotation. Hindered methyl rotation produces a characteristic orientation dependence of the relaxation times $T_{1Z}$ and $T_{1Q}$, which is easily distinguished from free rotation or small step rotational diffusion [129]. The activation energy for threefold methyl rotation is very sensitive to the local environment, and reflects different degrees of steric crowding. For example, the barrier is $4.0 \pm 0.3$ kJ/mol in acetone-$d_6$ [140], $2.9 \pm 0.5$ kJ/mol in copper acetate-$d_3$ monohydrate [138], and $2.6 \pm 0.5$ kJ/mol in acetanilide-$d_3$ [144]. For more hindered C-$(CD_3)_2$ groups larger values are typical; $19.2 \pm 2.0$ kJ/mol for bisphenol-A [145] and 15.3 and 22.2 kJ/mol for the two methyl groups of *N*-acetyl-DL-($\gamma$-$d_6$)-valine [146].

Deuteron spectra and spin lattice relaxation have been used by Polson et al. to look at the reorientation of *tert*-butyl groups in butylated hydroxytoluene [147]. They found motional inequivalence between the methyl groups within each *t*-Bu moiety: two methyl groups reorient rapidly ($C_3$) but the third has more restricted motion, reorienting at a rate comparable to the *t*-Bu group itself ($C_3'$). The temperature dependence of $T_{1Z}$ has two minima and gives activation energies of 9.25 kJ/mol and 25.2 kJ/mol for the two motions, respectively. Lineshape simulations of the *t*-Bu motion ($C_3'$) gave values close to those derived from the $T_{1Z}$ results. However, in order to account for the spectra above −80 °C it was necessary to postulate a third motion. Additional deuteron studies of complex motion of *tert*-butyl groups have been reported by Penner et al. [148] and Mooibroek et al. [149].

When the methyl spinning axis undergoes large angle motion, calculations of lineshapes and relaxation times can be sometimes be simplified by considering the motion of an effective quadrupole coupling tensor which has been averaged over the fast $C_{3v}$ motion. The largest principal component of the field gradient tensor is then directed along the spinning axis and the asymmetry parameter is zero. The validity of this approximation depends crucially on the motion of the spinning axis being much slower than the methyl rotation rate. If this is not true, multisite simulations must be used to provide accurate information about the rates of both processes.

### *4.1.2 Methylene Groups*

Deuterated methylene groups have been extensively used to study alkyl chain motion in liquid crystals [150–154], inclusion compounds [155–158], layered alkylammonium salts [159]; and cycloalkanes [160–162]. In all these investigations, in order to avoid ambiguous interpretations of motionally averaged spectra, it is important to establish reliable values for the quadrupole coupling tensor components in absence of the motion. Table 3 summarizes values of $CD_2$ quadrupole coupling tensors, determined for molecules in which no large amplitude motions occur.

In a single crystal study of deuterated malonic acid, Mueller et al. [3] found that the largest component of the methylene deuteron electric field gradient tensors was within 0.7° of the C—D bond direction. Two different QCC tensors were found for the methylene deuterons in 4-chlorophenyl acetic acid [163] and other substituted acetic acids [164]. This was attributed to the formation of a weak hydrogen bond with one of the deuterons. Two different methylene quadrupole tensors were also found in zero field measurements on diethylterephthalate-$d_{14}$ [60], but in this case the difference was a consequence of two crystallographically distinct environments. In malonic acid [164], both methylene deuterons were found to have the same QCC within experimental error (±0.2 kHz).

It is evident from the data collected in Table 3 that the values commonly adopted for typical methylene chain deuterons, QCC = 167 kHz and $\eta = 0$, are probably reliable within ±5 kHz and ±0.01, respectively. Thus, the observation of a small QCC (74 kHz) and large asymmetry parameter ($\eta = 0.36$) for solid nonadecane-2,2-$d_2$ [165] is a strong indication that large amplitude, restricted chain motion is occurring. Huang et al. showed that fast, restricted

**Table 3.** Methylene deuteron quadrupole coupling constants

| Substance | QCC | $\eta$ | Temperature | Method[1] | Reference |
|---|---|---|---|---|---|
| malonic acid | 165 | .03 | ambient | SC | 3 |
| α-glycine[2] | 169.4 | 0.085 | ambient | SC | 359 |
| α-glycine[2] | 160 | .043 | ambient | SC | 359 |
| ethanol[3] | 159.7 | 0.07 | 11 K | SC | 214 |
| ethanol[3] | 157.7 | 0.08 | 11 K | SC | 214 |
| ethanol[3] | 144.0 | 0.14 | 11 K | SC | 214 |
| ethanol[3] | 148.8 | 0.11 | 11 K | SC | 214 |
| diethylterephthalate[4] | 153.1 | .051 | ambient | ZF | 214 |
| diethylterephthalate[4] | 149.8 | .039 | ambient | ZF | 214 |
| cyclopentane | 171 | 0.02 | 83 K | QE | 160 |
| cyclohexane[5] | 166.7 | 0.00 | 127 K | QE | 219 |
| hexadecane[6] | 161.3 | 0.0 | 115 K | QE | 158 |
| hexadecane[6] | 80.3[7] | 0.0 | 298 K | SC | 157 |
| nonadecane[6] | 81.1[7] | 0.0 | 298 K | SC | 157 |
| icosane[6] | 81.9[7] | 0.0 | 298 K | SC | 157 |
| polyethylene | 164.9 | $<0.01$ | ambient | ZF | 59 |
| fluorene(9,9′) | 167.8 | 0.02 | ambient | SC | 360 |

[1] SC = single crystal; QE = quadrupole echo lineshape; ZF = zero field
[2] Different QCCs for each deuteron in the $CD_2$ group
[3] Four different sites for ethanol included in Dianin's compound
[4] Two crystallographically distinguishable sites
[5] Inclusion compound with thiourea
[6] Inclusion compound with urea
[7] Multiply by two to account for rapid rotation about the long molecular axis

motion of polymethylene chains could yield deuteron powder patterns with $\eta = 1.0$ [166], and Hirschinger and English consider alternative models which yield similar lineshapes [109].

### *4.1.3 Aromatic Deuterons*

Quadrupole coupling constants for aromatic deuterons (Table 4) are typically ~180 kHz, 10–20 kHz larger than those of aliphatic deuterons. Differences arising from substituent position are usually too small to detect by standard quadrupole echo experiments on perdeuterated materials, and are best investigated using single crystal rotation studies or zero field NMR on selectively deuterated compounds. Asymmetric, small angle libration of aromatic rings leads to motionally induced asymmetry parameters on the order of 0.02–0.10 [167], but this is difficult to distinguish from an intrinsic electronic asymmetry of the field gradient tensor. Armstrong et al. [168] have emphasized the importance of considering both origins of asymmetry in the analysis of motionally narrowed lineshapes.

For rigid molecules aromatic deuteron relaxation times are often painfully long. For example, quadrupole echo measurements in crystalline biphenyl-$d_{10}$ required 300 s recycle delays [167], and in dimethoxybenzene-$d_{10}$ $T_{1Z}$ was found to be 680 s [169]. Partially for this reason, systematic determinations of

**Table 4.** Aromatic deuteron quadrupole coupling constants

| Substance | QCC | $\eta$ | Temperature | Method[1] | Reference |
|---|---|---|---|---|---|
| benzene | 183 | 0.04 | 87 K | QE | 108 |
| toluene | 179.9 | 0.056 | 77 K | CW | 356 |
| naphthalene | 176.9 | 0.053 | 77 K | CW | 356 |
| 1,4-dibromobenzene | 176 | 0.09 | 298 K | MAS | 45 |
| terephthalic acid | 177.3 | 0.03 | ambient | ZF | 60 |
| diethylterephthalate[2] | 180.5 | 0.02 | ambient | ZF | 60 |
| diethylterephthalate[2] | 178.3 | 0.01 | ambient | ZF | 60 |
| *o*-benzoic acid | 177.8 | 0.04 | ambient | ZF | 60 |
| *m*-benzoic acid | 179.7 | 0.03 | ambient | ZF | 60 |
| *p*-benzoic acid | 172.9 | 0.01 | ambient | ZF | 60 |
| toluic acid[2] | 173.7 | 0.03 | ambient | ZF | 60 |
| toluic acid[2] | 178.9 | 0.02 | ambient | ZF | 60 |
| fluorene (1,4,5,8) | 178.8 | 0.047 | ambient | SC | 360 |
| fluorene (2,7) | 178.9 | 0.047 | ambient | SC | 360 |
| fluorene(3,6) | 172.9 | 0.047 | ambient | SC | 360 |

[1] CW = continuous wave (field or frequency sweep); QE = quadrupole echo lineshape; MAS = magic angle spinning; ZF = zero field NMR; SC = single crystal
[2] Crystallographically distinct sites

quadrupole coupling constants for chemically related aromatic compounds are very scarce. In aromatic compounds containing other faster relaxing deuterons, some relief from long relaxation times is afforded by MAS experiments, since this promotes relaxation by reintroducing spin diffusion [169]. The MAS technique appears to be a very promising way to investigate systems containing rigid aromatic rings.

In crystalline solids with sufficiently symmetric environments (near perfect hexagonal symmetry) aromatic rings undergo rapid rotation about the ring normal. Benzene-$d_6$ is the prototypical example and, as shown in Fig. 14, temperature dependent lineshape studies have been used to reinvestigate the kinetics. Similar $C_6$ motion was observed for benzene trapped in a cyclameric inclusion compound [108]. In hexamethylbenzene rotation about the ring normal has been investigated at low temperatures using quadrupole echo lineshapes [135], $T_{1Z}$ anisotropy [134], and 2D exchange methods [82]. Asymmetric substitution on the benzene ring completely quenches this $C_6$ motion, a result which points to the dominant role played by symmetry of the intermolecular potential in determining the barrier to rotation.

Aromatic rings attached to other molecular fragments, such as polymer chains [170] and intact enzymes [171–173], often execute large angle jumps about an axis parallel to the attachment axis. Simulations support the description of this motion as sudden discrete jumps between two sites, related by a 180° flip about an axis in the ring plane. There can be little doubt that this interpretation is basically correct, but there are in fact only a few investigations in which the flip angle has actually been determined from experimental data. For polycarbonate the 2D exchange experiment (shown in Fig. 11) allowed the flip angle distribution to be obtained. For polymers the ring motion is invariably characterized by a distribution of flip angles and jump rates [89], and simulations based on simple two site motion must be treated with caution.

### *4.1.4 N—D Groups*

Amine and amide groups are important to investigations of hydrogen bonding, and studies of their structure and dynamics are particularly relevant to peptides and proteins. QCC values for $-ND_2$ and $-ND_3^+$ deuterons are generally somewhat larger than those of deuterons bonded to carbon, and asymmetry parameters are typically in the range 0.1–0.2.

In the first single crystal $^2$H-NMR study of an amino acid, Usha et al. [174] found that the amide deuterons in *N*-acetyl glycine could be distinguished from carboxyl deuterons on the basis of extra line broadening caused by the nearby $^{14}$N. The largest component of the amide $^2$H EFG tensor was found to lie along the N—D bond direction (within experimental error) and the QCC values for N-acetyl glycine and L-pyroglutamic acid were found to be consistent with semi empirical relations [175, 176] between quadrupole coupling constants and N—D···O bond lengths. Usha et al. [174] emphasized the importance of using neutron diffraction rather than X-ray data as the source of bond lengths for such correlations. Very similar results have been reported for the amide quadrupole coupling parameters in *N*-acetyl-D,L-valine [54]. The relative orientation of the deuteron EFG and chemical shielding tensors was determined, and second order quadrupole shifts were analyzed. These two studies illustrate the wealth of accurate structural information which is the reward for extensive, time consuming, rotation studies of single crystals.

Most studies of N—D powder patterns have focused more on dynamics than structure [137, 177–183]. Quadrupole echo lineshapes of $-ND_3^+$ groups in simple molecular crystals have been simulated successfully with three-site jump models [137, 178]. Molecular motion in phenylethyl ammonium bromide [181] and phenylpropylammonium bromide [182] is more complicated, but the combination of quadrupole echo lineshapes, $T_{1Z}$ anisotropy, and $T_{1Q}$ measurements permitted the rates of both two- and three-fold jumps to be determined in these materials. Motion of $-ND_2$ groups is well characterized by two site, 180° flips. Kennedy et al. [183] noted that the standard procedure of simulating the pulse spacing dependence of quadrupole echo lineshapes fails for amide deuterons in *p*-nitroaniline, because $T_2$ is dominated by partially collapsed dipolar coupling to directly bonded $^{14}$N. $T_{1Z}$ anisotropy, which is unaffected by this process, was found to be a useful source of additional information. Williams and McDermott [100] combined measurements of quadrupole echo lineshapes and $T_{1Z}$ anisotropy to determine the activation energy for rotation about the C=O bond in pure crystalline urea. They found that the activation energy ($65 \pm 6$ kJ/mol) was three times larger than that measured in an urea/alkane inclusion compound [184]. Selective inversion was used to measure the rate of C—N rotation, in both urea and acetamide [100]. This technique was useful because the short transverse relaxation times precluded the use of 2D exchange, and C—N rotation was too slow to affect the lineshape directly.

#### *4.1.5 O—D Groups*

Quadrupole coupling parameters for hydroxyl deuterons are generally similar to those for deuterons bonded to nitrogen. Single crystal studies [3, 54, 185, 186] support the general conclusion that the largest field gradient is directed along the O—D bond. An extensive literature is devoted to correlations between hydroxyl deuteron quadrupole coupling constants and other measures of hydrogen bond strength, including bond distance [176, 187], infrared stretching frequencies [188], proton shielding tensors [189], and $C_\alpha$-$C_{acid}$-O torsion angle [164].

Deuteron NMR is the method of choice for investigating hydrogen bond dynamics. Benz et al. [185] analyzed the temperature and orientation dependence of deuteron $T_{1Z}$ in a $KDC0_3$ single crystal, using a model based on two site exchange in a double minimum potential with unequal minima. Early work in this area has been summarized by Scheubel et al. [190], who note that activation energies for hydrogen motion in H-bonded dimers fall into two classes. Low barriers ($\leqq 6$ kJ/mol) are associated with concerted jumps along the O—D···O bonds, while high barriers involve a two step process consisting of a 180° flip of the dimeric unit followed by fast deuteron rearrangement. The two step mechanism was found for dimethylmalonic acid. Scheubel et al. [190] noted that while their deuteron, carbon and proton data were all consistent with this description, $^{17}O$ data would be required for unambiguous proof of the models' validity.

Quadrupole echo lineshape analysis provided conclusive evidence for tetrahedral jumps in polycrystalline ice $I_h$ [191]. The activation energy for this motion, $55.7 \pm 0.8$ kJ/mol, is equal (within experimental error) to that of the Debye correlation time determined from dielectric relaxation. Independent measurements of deuteron spin alignment on doped ice samples [192] are in general agreement with these results, but reveal the existence of two distinct jump motions apparently associated with different types of Bjerrum defects. Molecular motion in solid and liquid phases of deuterated hydrogen sulfide and hydrogen selenide has been studied using temperature dependent QE lineshapes [193].

### *4.2 Disordered Materials and Surfaces*

Recent applications of deuteron NMR to amorphous and partly crystalline polymers have been reviewed by Spiess and coworkers [1, 2, 84], and therefore attention is confined in this section to studies of other classes of disordered materials: surfaces, glasses, and fibers.

Armstrong et al. analyzed temperature dependent quadrupole echo lineshapes of benzene-$d_6$ and pyridine-$d_5$ adsorbed on high surface area (140–155 $m^2/g$) $\gamma$-molybdenum nitride [194]. Chemisorbed benzene lies flat on metallic surfaces, but since pyridine can interact with the surface both through its aromatic $\pi$-electrons and the nitrogen lone pair, tilted or upright geometries for the adsorbate are not unreasonable. Two types of bound pyridine were observed in temperature programmed desorption studies [194]: a mobile fraction and a static fraction whose intensity decreased reversibly with increasing temperature. Lineshapes of the mobile fraction were successfully reproduced by a model of fast, incomplete, continuous rotation about the pyridine $C_{2V}$ axis, in an arc whose length decreased with increasing temperature. This was interpreted in terms of cooperative motion of clusters of tilted pyridine molecules.

Boddenburg and coworkers have investigated the orientation and rotational and translational dynamics of several small molecules adsorbed on graphite surfaces [195–199]. It was shown that anisotropic shielding, originating from the highly anisotropic solid support, competes for dominance with the electric quadrupole coupling. For neopentane [195], relaxation by chemical shift anisotropy was found to be dominant at temperatures just below the bulk melting point, and surface diffusion over spherical particles produced rapid, liquid like isotropic reorientation. Similar results were reported by Neue [200] for anthracene adsorbed at a cyclohexane/graphite interface.

Quadrupole echo lineshapes have been used to study the structure and dynamics of modified silica surfaces [201, 202]. Considerable differences in mobility of short chain adsorbates were found, depending on the extent of surface coverage [201]. The spectra for long chain adsorbates [202] at temperatures below ambient appear to be superpositions of two lineshapes, indicating that some silane groups move with a simple two site motion while others move with a more complex distribution of motions. These studies demonstrate the utility of $^2$H-NMR for studying surface mobility, but they also show that low S/N ratios restricts this method to systems with large surface area.

Vieth and coworkers used deuteron NMR to investigate the mobility of small deuterated molecules trapped in organic glasses [203, 204]. Relaxation times of spins on the rigid, glass forming matrix were too long to permit direct observation. However, the distribution of relaxation times and activation energies for rotationally mobile guests, such as hexamethylbenzene-$d_{18}$, provided information about local heterogeneity in the glass. Eckert and coworkers [187] investigated the structure and dynamics of water in silicate glasses. They successfully simulated QE lineshapes as a superposition of a broad distribution of quadrupole coupling constants, which was constructed from a detailed correlation between QCC values and the O—D—O bond distance. This and other, multinuclear NMR studies of disordered solids has been reviewed [205].

Drobny and coworkers have carried out extensive deuteron NMR studies of restricted motion in short DNA fragments [206–209]. These involved labeling the self complementary dodecamer $\{d(CGCGAATTCGCG)\}_2$ at various positions including methyl groups of dT7 and dT8, furanose rings dA5 and dA6, and the 5′,5″ positions of the 2′-deoxythymidine. By focusing attention on these short DNA fragments, complications due to imperfectly understood collective modes [210–213] are avoided. Quadrupole echo lineshapes, $T_{1Z}$ anisotropy, and $T_{2e}$ measurements were used to investigate the effects of increasing hydration on librational and overall motion. The data were best fit using a four site model of fast libration, and the libration amplitudes of all parts of the DNA fragment were found to increase slightly with increasing hydration. At high hydration levels, slow rotation of the whole dodecamer about the helix axis was observed.

## 4.3 Inclusion Compounds

There is considerable interest in both the structure and dynamics of the host and guest molecules forming this class of solids. Particular attention has been paid to what motions occur, and how they are modified by the constraining environment. These systems provide an ideal situation in which to realize the potential of $^2$H-NMR experiments to unravel detailed information on structure and dynamics. The investigation of ethanol in Dianin's compound, by Bernhard, Zimmerman and Haeberlen [214], was a tour de force and serves as an excellent example. Single crystals were grown of inclusion compounds containing three selectively deuterated ethanol guests; OD, $CD_2$ and $CD_3$. In order to understand the structures, extensive rotation studies were performed. The results were analyzed to obtain the eigenvalues (QCC and $\eta$) of the electric field gradient tensors (EFGs) at each deuteron position. At low temperatures (55 K), there are six sites available to the ethanol guests. Bond directions for all six sites were determined by invoking the assumption of tetrahedral bond angles. It was found that there are two molecules forming a weakly H-bonded dimer in each cage; one molecule is aligned roughly parallel to the crystallographic c axis (A), and the other (B) is approximately perpendicular. The bond directions are distributed around these average equilibrium directions. Even at low temperatures, $T = 11$ K, librational and torsional motions occur with amplitudes on the order of $\sim 10°$. The dynamics of the ethanol guests were investigated using temperature dependent spectra collected between 10 K–310 K. Spectra of deuterons jumping between distinct sites show characteristic lineshapes when the rate is comparable to the difference in quadrupolar splittings of the two sites (of the order of $10^5\,s^{-1}$). The ethanol molecules execute $C_3$ reorientations about the c axis at different rates for the two distinct orientations, A and B ($P^A(C_3) \neq P^B(C_3)$). Important additional information was deduced from two dimensional exchange experiments performed at a carefully chosen crystal orientation. These demonstrated that reorientations between the A and B molecules occur. Two types of AB reorientations were identified; the faster one ($k_{AB}^{(1)} = 200\,s^{-1}$ at 70 K, $\Delta E_a = 14$ kJ/mol) does not change the place of the two ethanol molecules within the cage, but the second one does ($k_{AB}^{(2)} = 10\,s^{-1}$ at 70 K). The authors suggest that the existence of the second type of guest reorientation implies that the host lattice of the clathrate is surprisingly flexible.

Zaborowski and Vega [215] have studied Dianin inclusion compounds with *p*- and *o*-Xylene guests. $^2$H-NMR spectra and $T_{1Z}$ relaxation times of partially deuterated guest and host molecules were measured as a function of temperature. The derived activation energies did not agree with predictions of molecular interaction minimum energy calculations. This result was rationalized by suggesting that the host molecules undergo dynamic processes that enhance the guests mobility.

Molecular motion in alkane/urea inclusion compounds has been investigated extensively by Vold et al. [155, 157, 184, 216–218]. These experiments focused on measuring the spectra and relaxation times of a series of different length alkanes in the urea channels. Individual spectral densities derived from the relaxation data were compared with results of Newtonian molecular dynamics simulations of alkane motion in a rigid host lattice. Although the dynamic simulations yielded spectral densities smaller than the experimental values by factors of 2–3, they indicated the possibility of minor populations of gauche conformations (especially near the ends of the alkane chains).

Recent results from a single crystal of urea-$d_4$/*n*-nonadecane inclusion compound [106] are shown in Figure 21. The urea molecules in this material undergo fast 180° flips about the C=O bond axis [184], and slower 180° flips about the C—N bond axes. The QE spectrum is shown for a particular orientation in Fig. 21a. Assignment of the lines is problematic because each urea molecule has two sets of inequivalent deuterons, (1, 2) and (3, 4), and can occupy any of six crystallographically distinct sites. The 2D exchange contour plot is shown for a mixing time of 25 ms in Fig. 21b. The C—N rotation results in cross peaks which identify deuterons in the same molecule, without recourse to time consuming rotation studies. In principle, the mixing time dependence of the cross peak intensity yields the C—N bond rotation rate, but long spin lattice relaxation times and limited signal to noise ratio conspire to make such experiments impractical. Instead, the rate $k_{CN}$ of C—N bond rotation was determined in a 1D experiment by monitoring the recovery of all transitions after selective inversion of peak 3.

Luz and coworkers have carried out detailed studies of several fluxional molecules trapped in urea and thiourea inclusion compounds [46, 219–221]. Thiourea can accommodate larger guests than urea: typically branched chain alkanes and aromatic molecules cannot be included in urea, while *tert*-butylbenzene readily forms an inclusion compound with thiourea [148]. Guest motion in thiourea channels can be quite complex. Cyclohexane-$d_{12}$ was found [219] to undergo several different whole body rotations in addition to chair-chair interconversion.

Ratcliffe et al. used deuteron NMR to investigate molecular motion in solid crown ether complexes [222]. For the malononitrile complex, 18-crown-6·$2CH_2(CN)_2$, motion was modeled as a combined rotation and conformational adjustment of the macrocycle, reminiscent of a merry go-round. Since this motion proceeds by sudden jumps between well defined potential minima, and does not induce disorder, it is invisible to X-ray or neutron diffraction measurements.

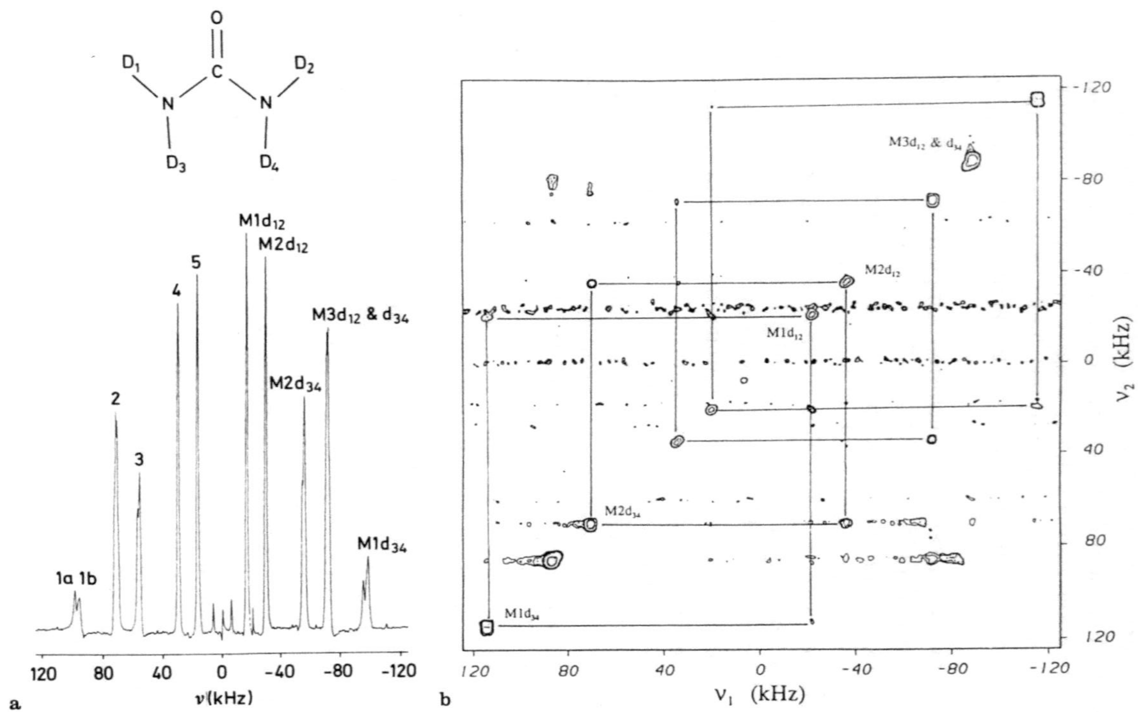


**Fig. 21a.** One dimensional QE spectrum of a single crystal urea-$d_4$/*n*-nonadecane inclusion compound, at an arbitrary orientation. The assignments were made with the aid of the 2D exchange spectrum shown in (**b**), for a mixing time of 25 ms. Cross peaks occur as a result of slow C—N rotation. The rate of this process can be determined and this allows identification of deuterons in the same molecule. The nomenclature is intended to distinguish equivalent pairs of deuterons, (*1*, *2*) and (*3*, *4*), from three crystallographically distinct molecules, M1, M2 and M3. Adapted with permission from Bell et al. [106]

In all inclusion compounds, the symmetry of both the guest and host cavities determine the available sites for jump type guest motion [223]. Temperature dependent distortions of the host lattice, which reduce this symmetry, may be partially responsible for the complex motional processes often found to occur in large cavities. In such cases, careful single crystal rotation studies are invaluable since they provide the most detailed kinetic information [214].

## 4.4 Liquid Crystals

### 4.4.1 Thermotropic

$^2$H-NMR has been extensively used to characterize ordering in thermotropic liquid crystals, and remains the preferred method for extracting orientational order parameters. One significant development is the increased attention being given to investigations of the liquid crystal molecules, rather than small rigid solutes dissolved in liquid crystalline solvents. In part, this is a result of the important advances that have been made in including the conformational dynamics of alkyl and alkoxy chains in theories of orientational order [224–231].

Quadrupolar splittings from deuterons attached to rigid subunits can be interpreted without recourse to models of internal rotation, and the goal of these studies is to determine the orientational order tensor [232–234]. When the number of tensor elements to be determined exactly matches the number of measured quadrupolar splittings it is always possible to obtain an exact fit to the data. However, accurate independent knowledge of the quadrupole coupling tensors (both magnitude and direction) is needed. For suitably symmetric systems with more than three deuterated sites, the problem of deriving the order tensor from measured splittings is overdetermined [234]. Compared with results based on one or two labeled positions this leads to an increased apparent error in the order parameters. However, the results are self consistent, less susceptible to systematic error and statistically significant. Such measurements are ideally suited to testing molecular theories of orientational order. For example, a new mean field description of nematic binary mixtures comprised of biaxial molecules [235, 236] has been explored using $^2$H-NMR data from 2-fluorenyl-4′-tetradecyloxybenzoate-$d_9$ ($FLOC_{14}$-$d_9$) and *para*-Xylene-$d_{10}$ [236, 237]. As shown in Fig. 22, the orientational order parameters, $S_{zz}$ and $S_{xx}-S_{yy}$, of the two components of the homogeneous mixture are very different. Ratios of the anisotropic molecular interaction potentials were derived by fitting the temperature dependence of experimental order parameters to theoretical expressions.

Mechanisms responsible for orienting molecules in anisotropic media have been explored in magic mixtures [231, 238]. These are liquid crystal mixtures where experiments on dissolved molecular hydrogen indicate that there is no external electric field gradient, thus only steric interactions are responsible for orienting molecules [239]. Models of this short range potential have been

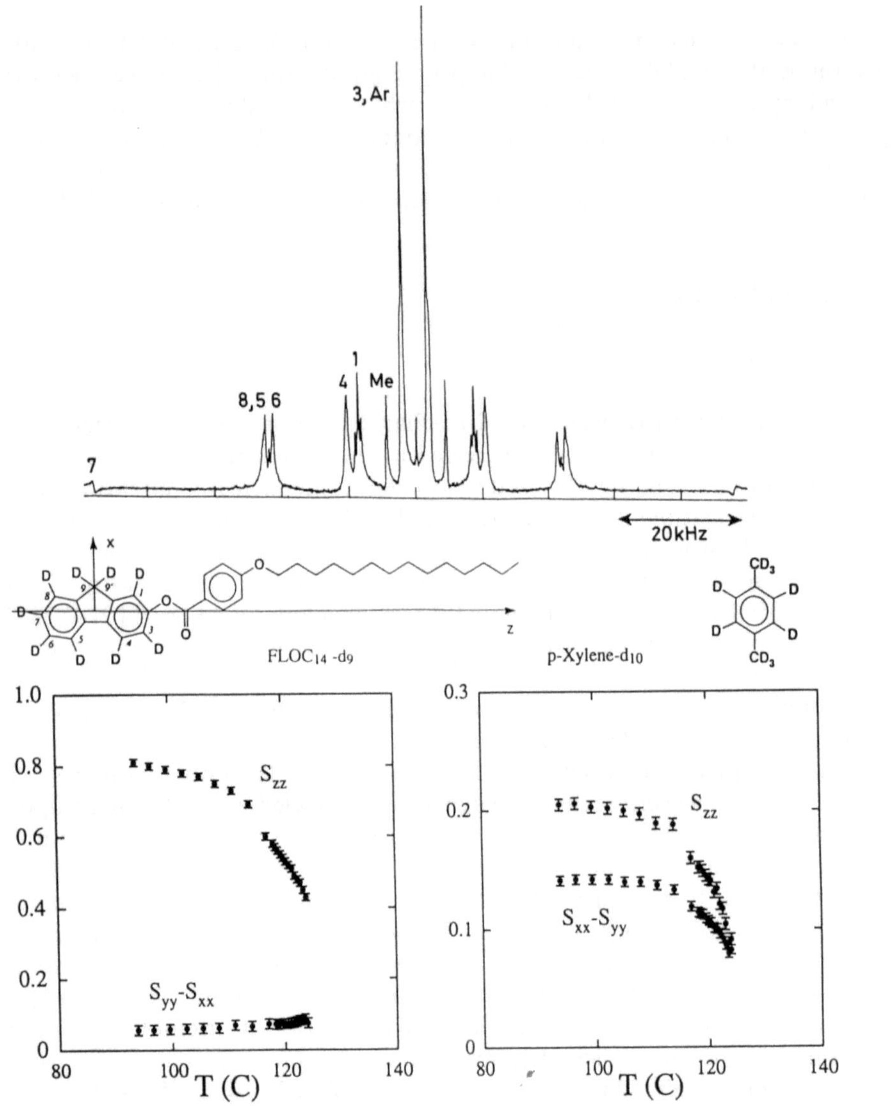


**Fig. 22.** This shows the QE spectrum from a nematic binary mixture comprised of the biaxial molecules 2-fluorenyl-4′-tetradecyloxybenzoate-$d_9$ (FLOC$_{14}$-$d_9$) and *para*-Xylene-$d_{10}$. Analysis of the temperature dependent quadrupolar splittings, from all labeled positions, yields the orientational order parameters, $S_{zz}$ and $S_{xx}-S_{yy}$, of both species. Adapted with permission from Hoatson et al. [237]

successful in describing and predicting the orientation of solute molecules of a variety of symmetries [226, 227, 238–241]. These models of orientational order have been extended to describe liquid crystal molecules and this requires explicit inclusion of internal rotation [231]. For 4-cyano-4′-*n*-pentylbiphenyl (5CB-$d_{19}$) in magic mixtures the observed quadrupolar splittings have been used to compare and fine tune two different models for the short range potential [240].

An alternative mean field description of conformational probabilities of alkyl chains in the presence of an orienting potential has been proposed by Photinos et al. [229, 230]. These two different theories [226, 227, 229–231] have been compared using both observed quadrupolar splittings and $^1$H dipolar couplings measured from double quantum filtered COSY experiments on a series of *n*-alkanes dissolved in liquid crystalline solvents [242–244]. Emsley and co-workers [225, 245–249] analyzed deuteron and proton NMR spectra of several solutes and nematogens to obtain information about potential functions for internal rotation in ordered media. A strong DC electric field has been used to probe pretransitional behavior of selectively deuterated 5CB [250].

Deuteron NMR is a standard method for investigating molecular motion in anisotropic fluids. Here the motivation is to determine the individual spectral densities of motion and to use these experimental results to assess the validity of theoretical descriptions of motion in an ordering potential. Recent advances have improved the accuracy with which the individual spectral densities can be determined, and hence improved the fidelity of the derived motional correlation times. These involve using broadband Jeener-Broekaert (BBJB) experiments, instead of the conventional frequency selective Jeener-Broekaert (JB), to determine $T_{1Q}$ [133]. For polycrystalline powders this has been discussed (Sect. 3.7) and illustrated (Fig. 19). For solutes in liquid crystals with only a few deuterated sites the results are not dramatically different, however as the number of quadrupolar doublets increase the BBJB method is the only sensible way of collecting the data efficiently and obtaining accurate results [133]. This is because the BBJB experiment permits uniform excitation of quadrupolar order, and echo detection can be used to improve the quality of the partially relaxed spectra. Combining this with a separately optimized inversion recovery experiment with echo detection (IRQE) allows $T_{1Z}$ to be determined for each deuterated site. Together the BBJB/IRQE experiments were found to be less susceptible to systematic error and give more accurate values of the spectral densities $J_1(\omega_0)$ and $J_2(2\omega_0)$. The BBJB/IRQE combination will likely supersede the conventional method of collecting relaxation data and indeed, several recent papers have adopted this methodology [251, 252].

Knowledge of individual motional spectral densities for several different sites on a rigid subunit provides a wealth of information about rotational anisotropy. Many investigators have studied small, rigid solutes in order to avoid the complications associated with internal rotation [121, 253–255]. A recent example of the use of relaxation time measurements to study anisotropic reorientation of a liquid crystal is that of 2-fluorenyl-4′-tetradecyloxybenzoate-d$_9$ (FLOC$_{14}$-d$_9$) [256]. The salient result was the success of the third rate model of molecular motion; this agreed, with reasonable accuracy and consistency, with the measured values of the spectral densities, $J_1^{(i)}(\omega_0)$ and $J_2^{(i)}(2\omega_0)$, for all seven inequivalent deuterons, i. Earlier investigations of rotational anisotropy in liquid crystals were less conclusive because, with a limited number of deuterated sites, there was insufficient experimental data to rigorously test motional models.

One interesting point to emerge from these studies is the contrast between results obtained for solutes [121, 253–255] and liquid crystal molecules [256–265]. It appears that for nematogens the correlation time for motion around the director is two to three orders of magnitude longer than that of typical solutes. As a result, the implicit separation of time scales for rotational dynamics and collective order director fluctuations is suspect. Several different theoretical formulations [266, 267] suggest that when rotational motion and order director fluctuations occur on comparable time scales, cross terms effectively quench contributions from collective motion. This explains why there was no need to invoke order director fluctuations to account for measured values of $J_1(\omega_0)$ for the liquid crystal $FLOC_{14}$-$d_9$ [256]. This contradicts results from small solutes, where large ODF contributions must be explicitly included in order to account quantitatively for the measured spectral densities [121, 253–255]. This clarifies the long standing puzzle concerning contradictory results from low frequency field cycling techniques [268–270] and high field spectral density measurements on solutes.

Recent theoretical advances in the description of rotational motion in liquid crystals include the calculation of spectral densities for a fully anisotropic rigid rotor with a biaxial restoring potential [271], and the development of models which couple the conformational dynamics of the alkyl chain with the

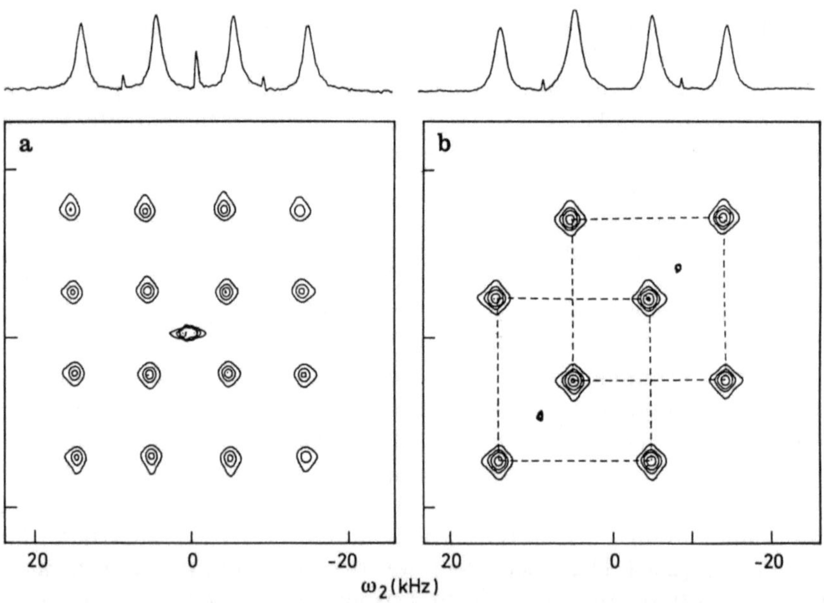


**Fig. 23a, b.** Deuteron 2D exchange spectra of cyclohexane-$d_{12}$ in nematic Phase V, at $-27\,°C$. The traces at the *top* of the contour plots are normal 1D QE spectra. The 2D spectra were obtained by (**a**) Zeeman order and (**b**) combined Zeeman and quadrupole order experiments, both with a pulse flip angle of $\theta = 54.7°$. The cross peaks arise as a result of intramolecular motion and their positions give the relative signs of the quadrupole interaction in the exchanging sites. Reproduced with permission from Boeffel et al. [93]

orientational diffusion of the molecule [272–275]. These models are less empirical than earlier models such as anisotropic viscosity [267, 276] or the third rate model [277]. Preliminary experimental tests appear promising [154], and these advances should allow interpretation of experimental data from a significantly broader range of liquid crystal systems.

The utility of deuteron 2D techniques for studying intramolecular exchange processes of solutes in oriented liquid crystalline solvents has been demonstrated by Luz and coworkers [92, 93, 117]. In a careful exposition Boeffel et al. [92] showed that, unlike exchange experiments on polycrystalline powders, in nematics additional phase cycling procedures are needed to remove spectral artifacts. Figure 23 illustrates the results obtained for axial-equatorial exchange of cyclohexane-$d_{12}$ in nematic Phase V at −27 °C. The 1D quadrupole echo spectra (top traces) show little evidence of the slow exchange process, while the 2D representation of (a) the stimulated echo (Zeeman order) experiment shows cross peaks connecting all the transitions. Exchange cross peaks appear exclusively as off diagonal signals in (b) the combined Zeeman and quadrupole order experiments. The cross peak positions determine the relative signs of the quadrupolar interactions in the exchanging sites, and the relative intensities of diagonal and off diagonal peaks provide estimates of the exchange rate. Thus, there is a close analogy between this 2D experiment and that used for exchange in isotropic liquids [278]. 2D exchange experiments in liquid crystals are especially promising because the scaled anisotropic interactions define a wider kinetic window than is accessible in isotropic fluids. However slow processes, such as order director fluctuations, which are unique to the ordered medium are probably not accessible because the angular extent of the fluctuations is too small.

There have been several papers describing liquid crystals droplets dispersed in polymer matrices [279–289]. Here the objective is to use $^2$H-NMR spectra to determine the static ordering of the liquid crystal in the droplet and at the polymer interface. The nematic director can adopt star, axial or bipolar configurations depending on the competition between the surface and the magnetic field in aligning the director within the droplets. 2D spectra and $T_{1Z}$ relaxation rates of 5CB-$d_2$, labeled in the 2 position of the alkyl chain, have been measured in the nematic and isotropic phases [286]. The results suggest a bipolar nematic director distribution in the droplets with a random distribution of their symmetry axes in the magnetic field. The existence of an orientationally ordered surface layer above the nematic-isotropic transition temperature was demonstrated [290]. Significant differences in the ratio ($T_{1Z}/T_2$) were found between the isotropic and nematic droplets and this was attributed to translationally induced rotation in the non uniformly oriented director field inside the cavity.

Deuteron NMR has been used in several recent studies of discotic liquid crystals [111, 253, 291–299]. In general, the motivation for studying and techniques used for discotic phases are identical to those of their calamitic counterparts.

### 4.4.2 Lyotropics

$^2$H-NMR is frequently used to elucidate the phase diagrams, determine details of molecular order, and characterize molecular motion in surfactant systems [300–310], microemulsion droplets [311–313] and ribbon bilayer phases [314, 315]. These systems are sufficiently complex that results based on any one spectroscopic technique probably provide an incomplete picture. In these investigations it is usual and advisable to combine deuteron NMR measurements with other physical techniques; multinuclear NMR spectroscopy ($^1$H, $^{13}$C, $^{31}$P or $^{23}$Na), diffusion coefficient measurements, differential scanning calorimetry, and X-ray or neutron scattering.

Deuteron NMR has become a standard method for investigating the structure and dynamics of membranes, phospholipid bilayers and related model systems. There are a few studies of orientational ordering of alcohols in lipid bilayers [316–318], but more effort has been devoted to understanding ordering and dynamics of acyl chains [319–322] and head groups [146, 323–326]. $^2$H-NMR structural studies of phospholipids often measure the quadrupolar splittings as a function of position on the alkyl chains [319, 320, 327]. These order parameter profiles can be obtained either from macroscopically oriented samples or from superimposed powder patterns by numerical deconvolution [328, 329].

Dynamic studies have focused on trying to characterize the extremely complicated spectrum of motional processes. Measurements of $T_{1Z}$ anisotropy [127, 330, 331] and more recently, the orientation dependence of individual spectral densities [251], have proven useful in settling the controversy over contributions from collective motions to deuteron spin lattice relaxation rates [330, 332–334]. It now appears that a combination of vesicle tumbling, lateral diffusion, and restricted wobbling relative to the bilayer normal adequately accounts for deuteron $T_{1Z}$ and $T_{1Q}$ data from 1–100 MHz, without contributions from collective motion [335]. However, 2D deuteron NMR studies of glycolipid gel phases suggest that whole molecule, large angle jumps occur with correlation times on the order of milliseconds [336–339].

Effects of cholesterol [209, 340–345], peptides and proteins [327, 346–349], and other additives [324, 340, 350–352] on the dynamics of phospholipid membranes have been intensively investigated using a variety of deuteron NMR techniques. Siminovitch et al. comment that lineshape simulations alone may be subject to erroneous interpretation, and should be supplemented by measurements of $T_{1Z}$ anisotropy [331]. Unfortunately, in some liposome systems the relaxation time anisotropy is washed out by fast translational diffusion over the curved membrane surface. Addition of cholesterol to many of these membranes restores the $T_{1Z}$ anisotropy, either by reducing the diffusion rate by an order of magnitude, or by markedly increasing the radius of curvature of the liposomes [331, 353].

The advantage of working with macroscopically oriented bilayers is illustrated by the investigation of Trouard et al. [251], who determined $T_{1Z}$ and $T_{1Q}$

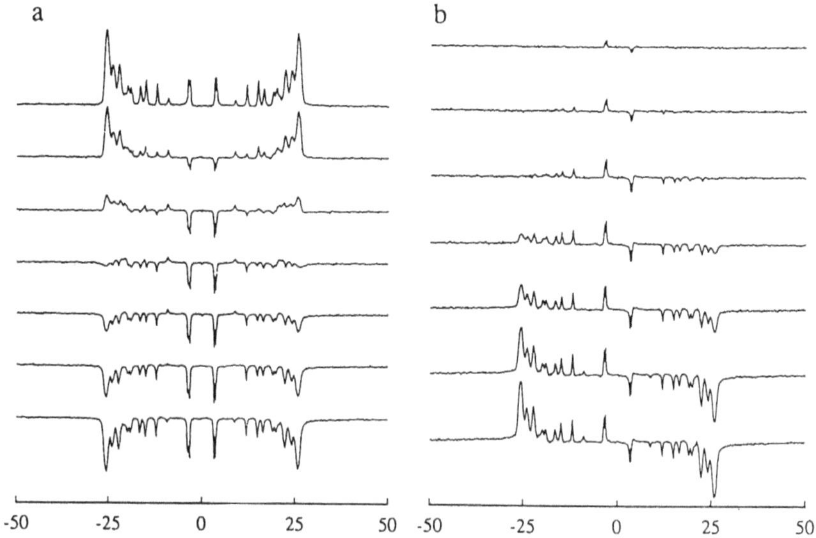


**Fig. 24a, b.** Partially relaxed $^2$H-NMR spectra of macroscopically oriented bilayers of DMPC-$d_{54}$: cholesterol (1:1 mole ratio) at 40 °C. These were obtained from the (**a**) inversion recovery quadrupolar echo sequence and (**b**) the broadband Jeener-Broekaert sequence. Reproduced with permission from Trouard et al. [251]

of DMPC-$d_{54}$ and cholesterol (1:1 mole ratio). The partially relaxed spectra are shown as stack plots in Fig. 24. The individual spectral densities $J_1(\omega_0)$ and $J_2(2\omega_0)$ were determined as a function of position on the acyl chain. These data could be explained using a simple model of rotational diffusion for the membrane lipid constituents. However, molecular motion in lipid membranes occurs on many different timescales and a combination of techniques is needed to provide a complete picture.

Investigations by Kothe and coworkers provide an excellent example of the detailed information which can be obtained by analyzing deuteron relaxation data from experiments which are sensitive to both slow and fast motional processes [126, 348, 354]. Measurements of $T_{1Z}$, $T_{1Q}$, $T_{2e}$, and 1D and 2D lineshapes were performed on selectively deuterated 1,2-dimyristoyl-sn-glycero-3-phosphocholine (DMPC) bilayers in unoriented and macroscopically aligned samples, as a function of temperature and orientation [126]. The results were fitted to a model of molecular motion based on the stochastic Liouville equation [355]. Intramolecular motion, consisting of *gauche-trans* isomerization of the individual segments of the phospholipid chain, which was represented by a jump process between distinct conformations at rate $\tau_J$. Intermolecular motion corresponding to restricted rotational diffusion of the molecule as a whole was characterized by correlation times for reorientation about axes parallel and perpendicular to the diffusion tensor axis. The uniaxial order parameter, $S_{ZZ}$, was used to characterize the ordering potential. In addition to the motion of isolated molecules, collective fluctuations (ODF) were incorporated in the model as changes of the instantaneous director with respect to its time averaged orientation, and a broad distribution of thermally activated hydrodynamic modes was invoked to provide information on viscoelastic properties of the bilayer membranes. Above the gel-liquid crystalline phase transition, cholesterol was found to increase the orientational and conformational order of the lipids. *gauche-trans* Isomerization rates of all chain segments are slowed down to values characteristic of the gel phase but molecular reorientation remained fast on the NMR timescale. At high temperatures, the ODF dominated the transverse relaxation rate. Cholesterol was found to reduce the density of the collective fluctuations by increasing the average elastic constant of the membrane. Although this model involves a large number of adjustable parameters, it appears that they can all be obtained with reasonable precision because the experiments are sensitive to a sufficiently wide range of timescales.

## 5 Conclusions

Solid state deuteron NMR spectroscopy is currently a mature yet vibrant field. Its extensive application to a diverse range of chemically, physically, and biologically interesting systems is a tribute to its versatility. Standard techniques, such as quadrupole echo lineshape analysis and measurements of spin lattice relaxation times, continue to provide valuable insights into ordering and motion at the molecular level. Although well established, the area continues to develop and expand. During the last five years new techniques have been introduced and perfected: multidimensional exchange experiments, measurements of individual spectral densities, and magic angle spinning. These have proven remarkably effective in extending the range of problems, systems and time scales which can be investigated using $^2$H-NMR. It is anticipated that deuteron NMR spectroscopy will continue to evolve, and thus remain a powerful probe of structure and molecular dynamics in solids and anisotropic fluids.

*Acknowledgments.* We are grateful to all authors who provided manuscripts prior to publication or gave us permission to reproduce figures from their work. Marco Brown and April Baugher are thanked for their help in making Fig. 13 and 16. We appreciate the hospitality extended to us by Alex Pines during our sabbatical visit to the University of California, Berkeley. He and the members of his research group provided a stimulating environment in which to complete this review. This work was supported in part by grant CHE-930557 from the National Science Foundation.

# 6 References


1. Spiess HW (1991) Chem Rev 91: 1321
2. Schmidt-Rohr K, Spiess HW (1994) Multidimensional solid state NMR and polymers. Academic, London
3. Mueller C, Izdiak S, Pislewski N, Haeberlen U (1982) J Magn Reson 47: 227
4. Rose ME (1957) Elementary theory of angular momentum. Wiley, New York
5. ter Beek LC, Burnell EE (1993) Bull Magn Reson 15: 175
6. Emsley JW, Lindon LC (1975) NMR Spectroscopy using liquid crystal solvents, Pergammon, New York
7. Heaton NJ, Vold RR, Vold RL (1989) J Chem Phys 91: 56
8. Haeberlen U (1976) High resolution NMR in solids: Selective averaging, Academic, New York
9. Torgeson DR, Schoenberger RJ, Barnes RG (1986) J Magn Reson 68: 85
10. Siminovitch DJ, Rance M, Jeffrey KR (1984) J Magn Reson 58: 62
11. Wittebort RJ (1989) J Magn Reson 83: 626
12. Cohen MH, Reif F (1957) Solid state physics 5: 321
13. Das TP (1958) Solid state physics 5: Supplement 1, 1
14. Mason J (1993) Solid state NMR 2: 285
15. Boden N, Clark LD, Hanlon SM, Mortimer M (1978) Faraday Symp Chem Soc 1979: 109
16. Davis JH, Jeffrey KR, Bloom M, Valic MI, Higgs TP (1976) Chem Phys Lett 42: 390
17. Spiess HW (1978) In: Diehl P, Fluck E, Kosfield R (eds) NMR: Basic principles and progress. Springer, Berlin, Heidelberg New York, vol 15, p 55
18. Bloom M, Davis JH, Valic MI (1980) Can J Chem 58: 1510
19. Griffin RG (1981) Methods in Enzymology 72: 108
20. Henrichs PM, Hewitt JM, Linder MJ (1985) J Magn Reson 60: 280
21. Spiess HW, Sillescu H (1981) J Magn Reson 42: 381
22. Hoatson GL, Packer KJ (1980) Mol Phys 40: 1153
23. Levitt MH, Suter D, Ernst RR (1984) J Chem Phys 80: 3064
24. Siminovitch DJ, Raleigh DP, Olejniczak ET, Griffin RG (1986) J Chem Phys 84: 2556
25. Heaton N, Vold RL, Vold RR (1988) J Magn Reson 77: 572
26. Raleigh DP, Olejniczak ET, Griffin RG (1989) J Magn Reson 81: 455
27. Wimperis S (1990) J Magn Reson 86: 46
28. Hoatson GL (1991) J Magn Reson 94: 152
29. Callahan J, Mattiello D, Drobny G (1992) (paper presented at ISMAR IX, Vancouver, B.C.)
30. Hoatson GL, Bluemich B (1991) J Magn Reson 95: 446
31. Bluemich B (1987) Prog NMR Spectr 19: 331
32. Paff J, Bluemich B, Kaiser R (1992) Adv Magn and Opt Reson 17: 1
33. Greferath M, Bluemich B, Griffith WM, Hoatson GL (1993) J Magn Reson A 102: 71
34. Scheler G, Hoatson GL, Bluemich B (1993) Private Communication
35. Ackerman JL, Eckman R, Pines A (1979) Chem Phys 42: 423
36. Maricq M, Waugh JS (1977) Chem Phys Lett 47: 327
37. Eckman R, Alla M, Pines A (1980) J Magn Reson 41: 440
38. Eckman R, Mueller L, Pines A (1980) Chem Phys Lett 74: 376
39. Eckman R, Pines A, Tycko R, Weitekamp DP (1983) Chem Phys Lett 99: 35
40. Vega S, Shattuck TW, Pines A (1976) Phys Rev Lett 37: 43
41. Vega S, Pines A (1977) J Chem Phys 66: 5624
42. Vega S, Shattuck TW, Pines A (1980) Phys Rev A 22: 638
43. Schadt RJ, Dong RY, Guenther E, Bluemich B (1992) J Magn Reson 96: 393
44. Kristensen JH, Bildsoe H, Jakobsen HJ, Nielsen NC (1992) J Magn Reson 100: 437
45. Kristensen JH, Bildsoe H, Jakobsen HJ, Nielsen NC (1991) J Magn Reson 92: 443
46. Reichert D, Olender Z, Poupko R, Zimmerman H, Luz Z (1993) J Chem Phys 98: 7699
47. Kustanovich I, Vega S, Zaboraowski E (1991) J Magn Reson 93: 441
48. Duer MJ, Levitt MH (1992) Solid State NMR 1: 211
49. Schmidt A, Vega S (1987) J Chem Phys 87: 6895
50. Schmidt A, Smith SO, Raleigh DP, Roberts JE, Griffin RG (1986) J Chem Phys 85: 4248
51. Bluemich B, Bluemler P, Jansen J (1992) Solid State NMR 1: 111
52. Ernst H, Freude D, Wolf I (1993) Chem Phys Lett 212: 588

53. Goldman M, Grandinetti PJ, Llor A, Olejniczak Z, Sachleben JR, Zwanziger JW (1992) J Chem Phys 97: 8947
54. Gerald R, Bernhard T, Haeberlen U, Rendell J, Opella S (1993) J Am Chem Soc 115: 777
55. Schmidt A, McKay RA, Schaefer J (1992) J Magn Reson 96: 644
56. Mueller L, Eckman R, Pines A (1980) Chem Phys Lett 76: 149
57. Zumbalydis N, O'Reilly JM, Landry CJT (1993) Paper presented at 34th ENC, St. Louis MO
58. Bielecki A, Zax DB, Zilm KW, Pines A (1986) Rev Sci Instru 57: 393
59. Millar JM, Thayer AM, Bielecki A, Zax DB, Pines A (1985) J Chem Phys 83: 934
60. Millar JM, Thayer AM, Zimmermann H, Pines A (1986) J Magn Reson 69: 243
61. Millar JM, Thayer AM, Zax DB, Pines A (1986) J Am Chem Soc 108: 5113
62. Thayer AM, Millar JM, Pines A (1986) Chem Phys Lett 129: 55
63. Noack F (1986) Prog NMR Spectr 18: 171
64. Koenig SH, Brown RDI (1990) Prog NMR Spectr 22: 487
65. Tycko R (1990) J Chem Phys 92: 5776
66. Yang Y, Schuster M, Bluemich B, Spiess HW (1987) Chem Phys Lett 139: 239
67. Mueller KT, Sun BQ, Chingas GC, Zwanziger JW, Terao T, Pines A (1990) J Magn Reson 86: 470
68. Wu Y, Sun BQ, Pines A (1990) J Magn Reson 89: 297
69. Mueller KT, Chingas GC, Pines A (1991) Rev Sci Instru 62: 1445
70. Mueller KT, Wooten EW, Pines A (1991) J Magn Reson 92: 620
71. Sun BQ, Baltisberger JH, Wu Y, Samosan A, Pines A (1992) Solid State NMR 1: 267
72. Eastman MA, Grandinetti PJ, Lee YK, Pines A (1992) J Magn Reson 98: 333
73. Chmelka BF, Mueller KT, Pines A, Stebbins J, Wu Y, Zwanziger JW (1989) Nature (London) 339: 42
74. Samoson A, Lippmaa E, Pines A (1988) Mol Phys 65: 1013
75. Frydman L, Chingas GC, Lee YK, Grandinetti PJ, Eastman MA, Barrall GA, Pines A (1992) J Chem Phys 97: 4800
76. Schmidt-Rohr K, Hehn M, Schaefer D, Spiess HW (1992) J Chem Phys 97: 2247
77. Chmelka BF, Schmidt-Rohr K, Spiess HW (1993) Macromol 26: 2226
78. Hagemeyer A, Brombacher L, Schmidt-Rohr K, Spiess HW (1990) Chem Phys Lett 167: 583
79. Zwanziger J (1993) (Paper Presented at 37th ACS Solid State NMR Symposium, Denver, CO)
80. Hoatson GL, Vold RL, Lee YK (1994) (unpublished data)
81. Vold RR, Vold RL (1991) Adv Magn and Opt Reson 16: 85
82. Schmidt C, Bluemich B, Wefing S, Kaufmann S, Spiess HW (1987) Ber Bunsenges Phys Chem 91: 1141
83. Schmidt C, Bluemich B, Spiess HW (1988) J Magn Reson 79: 269
84. Spiess HW (1991) J Noncryst Solids 131–133: 766
85. Wefing S, Kaufmann S, Spiess HW (1988) J Chem Phys 89: 1234
86. Wefing S, Spiess HW (1988) J Chem Phys 89: 1219
87. Zhang C, Liu Y, Chin YH, Jones AA, Inglefield PT, Shi J.-F, Dickenson CL, MacKnight WJ, Chien JC (1993) Macromol 26: 1008
88. Zhang C, Wang P, Jones AA, Inglefield PT, Kambour RP (1991) Macromol 24: 338
89. Hanson MT, Bluemich B, Boeffel C, Spiess HW, Morbitzer L, Zembrod A (1992) Macromol 25: 5542
90. Schmidt-Rohr K, Spiess HW (1990) Proc XXV Cong Ampere 512
91. Spiess HW, Schmidt-Rohr K (1992) Polymer Preprints Am Chem Soc Div Polym Chem 32
92. Boeffel C, Luz Z, Poupko R, Zimmermann H (1989) J Magn Reson 85: 329
93. Boeffel C, Luz Z, Poupko R, Vega AJ (1988) Israel J Chem 28: 238
94. De Jong AF, Kentgens APM, Veeman WS (1984) Chem Phys Lett 109: 337
95. Bloembergen N, Purcell EM, Pound RV (1948) Phys Rev 102: 679
96. Duncan TM, Thayer AM, Root TW (1990) J Chem Phys 92: 2663
97. Kuhns PL, Conradi MS (1982) J Chem Phys 77: 1771
98. Brown MJ, Hoatson GL, Vold RL (1992) Poster presented at 34th ENC, St. Louis, MO
99. Lin TH, Vold RR (1991) J Phys Chem 95: 9032
100. Williams JC, McDermott A (1993) J Phys Chem 97: 12393
101. Jonsen P (1990) Chem Phys Lett 170: 311
102. Abragam A (1961) Principles of Nuclear Magnetism, Clarendon Press, Oxford
103. Schmidt C, Wefing S, Bluemich B, Spiess HW (1986) Chem Phys Lett 130: 84
104. Warren W, Silver MS (1988) Adv Magn Reson 12: 248
105. Theimer D, Bodenhausen G (1992) Appl Magn Reson 3: 1071

106. Bell KA, Hoatson GL, Vold RL (1994) J Magn Reson A 108: 238
107. Griffin RG, Beshah K, Ebelhauser R, Huang TH, Olejniczak ET, Rice DM, Siminovitch DJ, Wittebort RJ (1988) NATO ASI C228: 81
108. Ok JH, Vold RR, Vold RL, Etter MC (1989) J Phys Chem 93: 7618
109. Hirschinger J, English AD (1989) J Magn Reson 85: 542
110. Hirschinger J, Miura H, Gardner KH, English AD (1990) Macromol 23: 2153
111. Hirschinger J, Kranig W, Spiess HW (1991) Coll Poly Sci 269: 993
112. Hoatson GL, Vold RL, Tse TY (1994) J Chem Phys 100: 4756
113. Mueller K, Schleicher A, Kothe G (1987) Mol Cryst Liq Cryst A153: 117
114. Schleicher A, Mueller K, Kothe G (1989) Liq Cryst 6: 489
115. Schleicher A, Muller K, Kothe G (1990) J Chem Phys 92: 6432
116. Ahmad SB, Packer KJ (1979) Mol Phys 37: 47
117. Mueller K, Poupko R, Luz Z (1991) J Magn Reson 93: 291
118. Vega AJ, Poupko R, Luz Z (1989) J Magn Reson 83: 111
119. Vold RL, Vold RR, Warner M (1988) Faraday Trans 84: 997
120. Reimer D, Heaton N, Schleicher A, Mueller K, Kothe G, Vilfan M (1994) J Chem Phys 100: 1693
121. Voight J, Jacobsen JP (1983) J Chem Phys 78: 693
122. Vold RR, Vold RL, Szeverenyi NM (1981) J Phys Chem 85: 1934
123. Vold RL, Vold RR, Poupko R, Bodenhausen G (1980) J Magn Reson 38: 141
124. Kontaxis G, Zieger G, Sterk H, Moskau D (1994) J Phys Chem 98: 420
125. Stohrer J, Groebner G, Reimer D, Weisz K, Mayer C, Kothe G (1991) J Chem Phys 95: 672
126. Weisz K, Groebner G, Mayer C, Stohrer J, Kothe G (1992) Biochem 31: 1100
127. Siminovitch DJ, Ruocco MJ, Olejniczak ET, Das Gupta SK, Griffin RG (1985) Chem Phys Lett 119: 251
128. Lin TH, Vold RR, Vold RL (1991) J Magn Reson 95: 71
129. Torchia DA, Szabo A (1982) J Magn Reson 49: 107
130. Wittebort RJ, Szabo A (1978) J Chem Phys 69: 1722
131. Jeener J, Broekaert P (1967) Phys Rev 157: 232
132. Spiess HW (1980) J Chem Phys 72: 6755
133. Hoatson GL, Tse TY, Vold RL (1992) J Magn Reson 98: 342
134. Tang J, Sterna L, Pines A (1980) J Magn Reson 41: 389
135. Meirovitch E, Freed JH (1979) Chem Phys Lett 64: 311
136. Idziak S, Haeberlen U, Zimmermann H (1991) Mol Phys 73: 571
137. Hiyama Y, Roy S, Guo K, Butler LG, Torchia DA (1987) J Am Chem Soc 109: 2525
138. Werner U, Mueller-Warmuth W (1993) Z Physik B91: 65
139. Lalowicz ZT, Werner U, Muller-Warmuth W (1988) Z Naturforsch 43A: 219
140. Werner U, Mueller-Warmuth W, Pines A (1990) Bull Magn Reson 12: 9
141. van Zijl PCM, Bothner-By AA (1988) J Magn Reson 79: 439
142. Hiyama Y (1990) Z Naturforsch 45A: 273
143. Wann MH, Harbison GS (1994) J Chem Phys 101: 231
144. Fletcher D, Hoatson GL, Vold RL (1993) (unpublished observations)
145. Tse TY, Hoatson GL, Vold RL (1993) (poster presented at ACS Polymer Symposium, Keystone, CO)
146. Beshah K, Griffin RG (1989) J Magn Reson 84: 268
147. Polsen JM, Fyfe JDD, Jeffrey KR (1990) J Chem Phys 94: 3381
148. Penner GH, Polson JM, Stuart C, Ferguson G, Kaitner B (1992) J Phys Chem 96: 5121
149. Mooibroek S, Wasylishen RE, Macdonald JB (1988) Can J Chem 66: 734
150. Barbara TM, Vold RR, Vold RL, Neubert ME (1985) J Chem Phys 82: 1612
151. Dong RY, Richards GM (1989) J Chem Phys 91: 7276
152. Dong RY, Richards GM (1990) Chem Phys Lett 171: 389
153. Dong RY (1991) Phys Rev A 43: 4310
154. Ferrarini A, Nordio PL, Moro GJ (1991) Mol Cryst Liq Cryst 198: 159
155. Greenfield MS, Vold RL, Vold RR (1989) Mol Phys 66: 269
156. Greenfield MS (1987) PhD Thesis, U.C. San Diego, La Jolla, CA
157. Vold RL, Vold RR, Heaton NJ (1989) Adv Magn Reson 13: 17
158. Harris KDM, Jonsen P (1989) Chem Phys Lett 154: 593
159. Jurga Ş, Macho V, Hueser B, Spiess HW (1991) Zh Physik B 84: 43
160. Lizak MJ, Keller RC, Coffey MS, Conradi MS (1990) J Phys Chem 94: 992
161. Mack JW, Torchia DA (1991) J Phys Chem 95: 4207
162. Schlick S, Luz Z, Poupko R, Zimmermann H (1992) J Am Chem Soc 114: 4315

163. Jackisch MA, Jarrett WL, Guo K, Fronczek FR, Butler LG (1988) J Am Chem Soc 110: 343
164. Jarrett WL, Guo K, Jackisch MA, Butler LG (1989) J Magn Reson 82: 76
165. Taylor MG, Kelusky EC, Smith ICP, Casal HC, Cameron DG (1983) J Chem Phys 78: 5108
166. Huang TH, Skarjune RP, Wittebort RJ, Griffin RG, Oldfield E (1980) J Am Chem Soc 102: 7377
167. Ronemus AD, Vold RL, Vold RR (1986) J Magn Reson 70: 416
168. Armstrong PA, Bell AT, Reimer JA (1993) Solid State NMR 2: 1
169. Alla M, Eckman R, Pines A (1980) Chem Phys Lett 71: 148
170. Roy AK, Inglefield PT (1990) Prog NMR Spectr 22: 569
171. Sparks SW, Cole HBR, Torchia DA (1989) Chem Scripta 29A: 31
172. Burke PA, Griffin RG, Klibanov AM (1992) J Biol Chem 267: 20057
173. Burke PA, Griffin RG, Klibanov AM (1993) Biotechnology and Bioengineering 42: 87
174. Usha MG, Peticolas WL, Wittebort RJ (1991) Biochem 30: 3955
175. Hunt MJ, Mackay AL (1976) J Magn Reson 22: 295
176. Looyestijn WJ, Klaassen TO, Poulis NJ (1979) Physica B96: 303
177. Grosescu R (1988) Revue Roumaine de Physique 33: 637
178. Szabo A, Torchia DA (1985) J Magn Reson 64: 134
179. Ikeda R (1989) J Phys Chem 93: 7315
180. Pratum TK (1989) J Magn Reson 81: 350
181. Gruwel MLH, Wasylishen RE (1991) Z Naturforsch 46A: 691
182. Grewel MLH, Wasylishen RE (1992) Z Naturforsch 47A: 1073
183. Kennedy MA, Vold RR, Vold RL (1991) J Magn Reson 91: 301
184. Heaton NJ, Vold RL, Vold RR (1989) J Am Chem Soc 111: 3211
185. Benz S, Haeberlen U, Tegenfeld J (1986) J Magn Reson 66: 125
186. Tesche B, Zimmerman H, Poupko R, Haeberlen U (1993) J Magn Reson A 104: 68
187. Eckert H, Yesnowski JP, Stolper EM (1989) Solid State Ionics 32: 298
188. Berglund B, Lindgren J, Tegenfeld J (1978) J of Mol Struct 43: 179
189. Bergland B, Vaughan RW (1980) J Chem Phys 73: 2037
190. Scheubel W, Zimmermann H, Haeberlen U (1988) J Magn Reson 80: 401
191. Wittebort RJ, Usha MG, Ruben DJ, Wemmer DE, Pines A (1988) J Am Chem Soc 110: 5668
192. Fujara F, Wefing S, Kuhs WH (1988) J Chem Phys 88: 6801
193. Collins MJ, Ratcliffe CI, Ripmeester JA (1989) J Phys Chem 93: 7495
194. Armstrong PA, Bell AT, Reimer JA (1993) J Phys Chem 97: 1952
195. Auer G, Boddenberg B (1992) J Magn Reson 100: 88
196. Boddenberg B, Grosse R (1987) Z Naturforsch 42A: 272
197. Boddenberg B, Grosse R (1988) Z Naturforsch 43A: 497
198. Boddenberg B, Grundke V (1991) Z Naturforsch 46A: 211
199. Horstmann W, Auer G, Boddenberg B (1987) Zeitschrift fur Physikalische Chemie, Neue Folge, 152: 23
200. Neue G (1988) Z Naturforsch 43A: 30
201. Kang H.-J, Blum FD (1991) J Phys Chem 95: 9391
202. Zeigler RC, Maciel GE (1991) J Am Chem Soc 113: 6349
203. Boerner F, Diezemann G, Roessler E, Vieth H.-M. (1991) Chem Phys Lett 181: 565
204. Rossler E, Taupitz M, Vieth H.-M (1989) Ber Bunsenges Phys Chem 93: 1241
205. Eckert H (1992) Prog NMR Spectr 24: 159
206. Alam TM, Drobny GP (1990) Biochem 29: 3421
207. Alam TM, Orban J, Drobny GP (1991) Biochem 30: 9229
208. Kintanar A, Huang W.-C, Schindele DC, Wemmer D, Drobny GP (1989) Biochem 28: 282
209. Huang W.-C, Orban J, Kintanar A, Reid BR, Drobny GP (1990) J Am Chem Soc 112: 9059
210. Brandes R, Vold RR, Vold RL, Kearns DR (1987) Springer Ser Biophys 1: 242
211. Vold RR, Brandes R, Tsang P, Kearns DR, Vold RL, Rupprecht A (1986) J Am Chem Soc 108: 302
212. Brandes R, Rupprecht A, Kearns DR (1989) Biophys J 56: 683
213. Tsang PT, Kearns DR, Vold RR (1992) J Am Chem Soc 114: 6585
214. Bernard T, Zimmerman H, Haeberlen U (1992) Mol Phys 77: 1123
215. Zaborowski E, Vega S (1993) Mol Phys 78: 703
216. Greenfield MS, Vold RL, Vold RR (1985) J Chem Phys 83: 1440
217. Heaton NJ, Vold RL, Vold RR (1989) J Magn Reson 84: 333
218. Cannarozzi GM, Meresi GH, Vold RL, Vold RR (1991) J Phys Chem 95: 1525
219. Poupko R, Furman E, Mueller K, Luz Z (1991) J Phys Chem 95: 407
220. Gelerinter E, Luz Z, Poupko R, Zimmermann H (1990) J Phys Chem 94: 5391

221. Gelerito F, Luz Z, Poupko R, Zimmemman H (1990) J Phys Chem 94: 5391
222. Ratcliffe CI, Ripmeester JA, Buchanan GW, Denike JK (1992) J Am Chem Soc 114: 3294
223. Meirovitch E, Rananavare SB, Freed JH (1987) J Phys Chem 91: 5014
224. Luckhurst GR (1988) J Chem Soc Faraday Trans II, 84: 961
225. Counsell CJR, Emsley JW, Luckhurst GR, Sachdev HS (1988) Mol Phys 63: 33
226. Van der Est AJ, Kok MY, Burnell EE (1987) Mol Phys 60: 397
227. Zimmerman DS, Burnell EE (1993) Mol Phys 78: 687
228. Photinos DJ, Samulski ET, Toriumi H (1992) Faraday Trans 88: 1875
229. Photinos DJ, Samulski ET, Toriumi H (1990) J Phys Chem 94: 4688
230. Photinos DJ, Samulski ET, Toriumi H (1990) J Phys Chem 94: 4694
231. Hoatson GL, Bailey AL, Van der Est AJ, Bates GS, Burnell EE (1988) Liq Cryst 3: 683
232. Diehl P, Ugolini R (1990) Mol Cryst Liq Cryst 188: 119
233. Bailey AL, Bates GS, Burnell EE, Hoatson GL (1989) Liq Cryst 5: 941
234. Wu G.-B, Ziemnicka B, Doane JW (1988) J Chem Phys 88: 1373
235. Palffy-Muhoray P, Hoatson GL (1991) Phys Rev A 44: 5052
236. Hoatson GL, Goetz JM, Palffy-Muhoray P, Crawford GP, Doane JW (1991) Mol Cryst Liq Cryst 203: 45
237. Hoatson GL, Goetz JM (1991) J Chem Phys 94: 3885
238. Zimmerman DS, Burnell EE (1990) Mol Phys 69: 1059
239. Barker PB, Van der Est AJ, Burnell EE, Patey GN, de Lange CA, Snijders JG (1984) Chem Phys Lett 107: 426
240. Chandrakumar T, Zimmerman DS, Bates GS, Burnell EE (1994) Liq Cryst (in press)
241. Kok MY, van der Est AJ, Burnell EE (1988) Liq Cryst 3: 485
242. Gochin M, Pines A, Rosen ME, Rucker SP, Schmidt C (1990) Mol Phys 69: 671
243. Gochin M, Hugi-Cleary D, Zimmermann H, Pines A (1987) Mol Phys 60: 205
244. Gochin M, Zimmermann H, Pines A (1987) Chem Phys Lett 137: 51
245. Emsley JW, Heaton NJ, Luckhurst GR, Shilstone GN (1988) Mol Phys 64: 377
246. Emsley JW (1989) Israel J Chem 28: 297
247. Celebre G, Longeri M, Emsley JW (1989) Liq Cryst 6: 689
248. Emsley JW, Luckhurst GR, Sachdev HS (1989) Liq Cryst 5: 953
249. Celebre G, Longeri M, Sicilia E, Emsley JW (1990) Liq Cryst 7: 731
250. Emsley JW, Imrie CT, Luckhurst GR, Newmark RD (1988) Mol Phys 63: 317
251. Trouard TP, Alam TM, Zajicek J, Brown MF (1992) Chem Phys Lett 189: 67
252. Dong RY (1992) Bull Magn Reson 14: 134
253. Joghems EA, Biesheuvel CA, Bulthuis J (1993) Liq Cryst 13: 427
254. Plomp L, Schreurs M, Bulthuis J (1988) J Chem Phys 88: 5202
255. Selwyn LS, Vold RL, Vold RR (1984) J Chem Phys 80: 5418
256. Goetz JM, Hoatson GL, Vold RL (1992) J Chem Phys 97: 1306
257. Dong RY, Richards GM (1988) Faraday Trans 84: 1053
258. Dong RY (1988) J Chem Phys 88: 3962
259. Dong RY, O'Bannon GW (1991) Mol Cryst Liq Cryst 209: 139
260. Dong RY (1990) Phys Rev A 42: 858
261. Dong RY, Emsley JW, Hamilton K (1989) Liq Cryst 5: 1019
262. Dong RY (1989) Liq Cryst 4: 505
263. Ikenberry DA (1989) PhD Thesis, Univ of Calif. San Diego
264. Beckmann PA, Emsley JW, Luckhurst GR, Turner DL (1983) Mol Phys 50: 699
265. Beckmann PA, Emsley JW, Luckhurst GR, Turner DL (1986) Mol Phys 59: 97
266. Gertner BJ, Lindenberg K (1991) J Chem Phys 94: 5143
267. Freed JH (1977) J Chem Phys 66: 4183
268. Koellner K, Schweikert H, Noack F (1993) Liq Cryst 4: 483
269. Noack F, Notter M, Weiss W (1988) Liq Cryst 6: 907
270. Scheufler P, Leiningen D, Woeffel D, Noack F (1988) Proceedings of the Colloque Ampere 22: 72
271. Tarroni R, Zannoni C (1991) J Chem Phys 95: 4550
272. Ferrarini A, Moro GJ, Nordio PL (1990) Liq Cryst 8: 593
273. Ferrarini A, Moro G, Nordio PL, Polimeno A (1988) Chem Phys Lett 151: 531
274. Ferrarini A, Nordio PL, Moro GJ, Crepeau R, Freed JH (1989) J Chem Phys 91: 5707
275. Ferrarini A, Moro GJ, Nordio PL (1992) Mol Phys 77: 1
276. Polnaszek CF, Freed JH (1975) J Phys Chem 79: 2283
277. Vold RR, and Vold RL (1988) J Chem Phys 88: 1443

278. Macura S, Ernst RR (1980) Mol Phys 41: 95
279. Golemme A, Zumer S, Allender DW, Doane JW (1988) Phys Rev Lett 61: 2937
280. Doane JW, Golemme A, West JL, Whitehead JB Jr, Wu BG (1988) Mol Cryst Liq Cryst 165: 511
281. Golemme A, Zumer S, Doane JW, Neubert ME (1988) Phys Rev A, 37: 559
282. Pirs J, Olenik M, Marin B, Zumer S, Doane JW (1990) J Appl Phys 68: 3826
283. Aloe R, Chidichimo G, Golemme A (1991) Mol Cryst Liq Cryst 203: 9
284. Allender DW, Crawford GP, Doane JW (1991) Phys Rev Lett 67: 1442
285. Crawford GP, Allender DW, Doane JW, Vilfan M, Vilfan I (1991) Phys Rev A 44: 2570
286. Dolinsek J, Jarh O, Vilfan M, Zumer S, Blinc R, Doane JW, Crawford G (1991) J Chem Phys 95: 2154
287. Stannarius R, Crawford GP, Chien LC, Doane JW (1991) J Appl Phys 70: 135
288. Crawford GP, Steele LM, Ondris-Crawford R, Iannacchione GS, Yeager CJ, Doane JW, Finotello D (1992) J Chem Phys 96: 7788
289. Crawford GP, Allender DW, Doane JW (1992) Phys Rev A, 45: 8693
290. Crawford GP, Stannarius R, Doane JW (1991) Phys Rev A, 44; 2558
291. Cheng GQ, Dong RY (1988) J Chem Phys 89: 3308
292. Vallerien SU, Werth M, Kremer F, Spiess HW (1990) Liq Cryst 8: 889
293. Hueser B, Pakula T, Spiess HW (1989) Macromol 22: 1960
294. Jolley KW, Smith MH (1989) Chem Phys Lett 162: 152
295. Kranig W, Boeffel C, Spiess HW (1990) Liq Cryst 8: 375
296. Kranig W, Huser B, Spiess HW (1990) Advanced Materials 2: 36
297. Richards GM, Dong RY (1989) Liq Cryst 5: 1011
298. Vallerien SU, Kremer F, Hueser B, Spiess HW (1989) Coll Poly Sci 267: 583
299. Abis L, Arrighi V, Cometti G, Dalcanale E, Du Vosel A (1991) Liq Cryst 9: 277
300. Boden N, Corne SA, Jolley KW (1987) J Phys Chem 91: 4092
301. Boden N, Clements J (1990) J Chem Phys 93: 9096
302. Blackmore ES, Tiddy GJT (1990) Liq Cryst 8: 131
303. Narayan KS, Shinde N, Tiddy GJT (1992) J Phys Chem 96: 5165
304. Chidichimo G, Golemme A, Ranieri GA, Terenzi M, Doane JW (1986) Mol Cryst Liq Cryst 132: 275
305. Chidichimo G, Golemme A, Doane JW, Westerman PW (1985) J Chem Phys 82: 536
306. Chidichimo G, Imbardelli D, Golemme A, Ranieri GA (1991) Mol Cryst Liq Cryst 201: 125
307. Chidichimo G, Coppola L, La Mesa C, Rainieri GA (1988) Chem Phys Lett 145: 85
308. Warnheim T (1987) J Coll Int Sci 118: 233
309. Warnheim T (1988) Liq Cryst 3: 791
310. Svard M, Schurtenberger P, Fontell K, Jonsson B, Lindman B (1988) J Phys Chem 92: 2261
311. Carlstrom G, Halle B (1988) Langmuir 4: 1346
312. Quist P.-O, Halle B (1988) J Chem Soc Faraday Trans. I, 84: 1033
313. Carlstrom G, Halle B (1989) J Phys Chem 93: 3287
314. Chidichimo G, Golemme A, Doane JW (1985) J Chem Phys 82: 4369
315. Raudino A, Grasso D, La Rosa C, Dipasquale G (1989) Liq Cryst 6: 435
316. Pope JM, Doane JW (1987) J Chem Phys 87: 3201
317. Pope JM, Westerman PW, Doane JW (1987) Chem Phys Lett 139: 589
318. Westerman PW, Pope JM, Phonphok N, Doane JW, Dubro DW (1988) Biochim Biophys Acta 939: 64
319. Sternin E, Fine B, Bloom M, Tilcock CPS, Wong KF, Cullis PR (1988) Biophys J 54: 689
320. Devaux PF, Hoatson GL, Favre E, Fellman P, Farren B, MacKay AL, Bloom M (1986) Biochem 25: 3804
321. Salmon A, Dodd S.W, Williams GD, Beach JM, Brown MF (1987) J Am Chem Soc 109: 2600
322. Jansson M, Thurmond RL, Barry JA, Brown MF (1992) J Phys Chem 96: 9532
323. Gutman H (1991) Liq Cryst 9: 607
324. Sanders CR, Prestegard JH (1990) Biophys J 58: 447
325. Delikatny EJ, Burnell EE (1989) Mol Phys 67: 757
326. Davies MA (1992) Biophys J 63: 1059
327. Nezil FA, Morrison C, Whittall KP, Bloom M (1991) J Magn Reson 93: 279
328. Sternin E, Bloom M, McKay AL (1983) J Magn Reson 55: 274
329. Bloom M, Davis JH, MacKay AL (1981) Chem Phys Lett 80: 198
330. Jarrell HC, Smith ICP (1988) J Chem Phys 88: 1260
331. Siminovitch DJ, Ruocco MJ, Olejniczak ET, Das Gupta SK, Griffin RG (1988) Biophys J 54: 373

332. Brown MF, Ellena JF, Trindle C, Williams GD (1986) J Chem Phys 84: 465
333. Brown MF, Salmon A (1990) Mol Phys 69: 379
334. Pastor RW, Venable RM, Karplus M, Szabo A (1988) J Chem Phys 89: 1128
335. Halle B (1991) J Phys Chem 95: 6724
336. Auger M, Jarrell HC (1990) Chem Phys Lett 165: 162
337. Winsborrow BG, Smith ICP, Jarrell HC (1991) Biophys J 59: 729
338. Auger M, Smith ICP, Jarrell HC (1991) Biophys J 59: 31
339. Winsborrow BG, Brisson J.-R, Smith ICP, Jarrell HC (1992) Biophys J 63: 428
340. Makriannis A, Yang DP, Griffin RG, Das Gupta SK (1990) Biochim Biophys Acta 1028: 31
341. Huang TH, Lee CWB, Das Gupta SK, Blume A, Griffin RG (1993) Biochem 32: 13277
342. Davis JH (1989) Adv Magn Reson 13: 195
343. Singh D, Davis JH, Grant CWM (1992) Biochim Biophys Acta 1107: 23
344. Vist MR, Davis JH (1990) Biochem 29: 451
345. Reinl H, Brumm T, Bayerl TM (1992) Biophys J 61: 1025
346. Prosser RS, Davis JH, Mayer C, Weisz K, Kothe G, (1992) Biochem 31: 9355
347. Lee CWB, Das Gupta SK, Mattai J, Shipley GG, Abdel-Mageed OH, Makriyannis A, Griffin RG (1989) Biochem 28: 5000
348. Meier P, Sachse JH, Brophy PJ, Marsh D, Kothe G (1987) Proc Natl Acad Sci USA, 84: 3704
349. Datema KP, van Boxtel BJH, Hemminga MA (1988) J Magn Reson 77: 372
350. Ulrich AS, Watts A (1993) Solid State NMR 2: 21
351. Blume A, Hubner W, Muller M, Bauerle HD (1988) Ber Bunsenges Phys Chem 92: 964
352. Huang TH, Blume A, Das Gupta SK, Griffin RG (1988) Biophys J 54: 173
353. Speyer JB, Weber RT, Das Gupta SK, Griffin RG (1989) Biochem 28: 9569
354. Mayer C, Mueller K, Weisz K, Kothe G (1988) Liq Cryst 3: 797
355. Mueller K, Meier P, Kothe G (1985) Prog NMR Spectr 17: 211
356. Barnes RG (1972) Adv Nuc Quad Reson 1: Ch 26
357. Rinne M, Depireaux J (1972) Adv Nuc Quad Reson 1: Ch 27
358. Zax DB, Bielecki A, Zilm KW, Pines A, Weitekamp DP (1985) J Chem Phys 83: 4877
359. Mueller C, Schajor W, Zimmerman H, Haeberlen U (1984) J Magn Reson 84: 235
360. Hoffmann W (1990) PhD Thesis, Free University of Berlin

# Cross-Polarization, Relaxation Times and Spin-Diffusion in Rotating Solids

D. Michel[1] and F. Engelke[2]

[1] Fakultät für Physik und Geowissenschaften der Universität Leipzig, Linnéstrasse 5, 04157 Leipzig, FRG
[2] Ames Laboratory, Iowa State University, Spedding Hall, Ames, Iowa 50011, USA

## Table of Contents



NMR Basic Principles and Progress, Vol. 32

The cross-polarization technique in conjunction with sample spinning has been widely applied in high-resolution solid-state NMR studies. The present article reviews the major work performed during the last decade concerning the basic physical background and experimental intricacies of cross-polarization in rotating solids. The main focus is directed towards Hartmann-Hahn cross-polarization and nuclear spin systems containing spin-1/2 nuclei. However, in the introductory part a brief account is given regarding studies on quadrupolar nuclei and NMR techniques employing cross-polarization schemes and sample spinning with these nuclear species.

As starting point, at first cross-polarization in non-rotating solids is considered in several levels of description: (i) the characterization of spin systems by their dipolar Hamiltonians, (ii) the straightforward thermodynamic model, (iii) the phenomenological description of cross-polarization dynamics, (iv) the statistical treatment of cross-polarization dynamics, and (v) the modified spin-thermodynamic theory to account for Hartmann-Hahn mismatch conditions.

Cross-polarization in rotating solids is characterized first by effects which sample spinning produces upon the Hartmann-Hahn matching curve: sample rotation leads to a break-up of the matching curve into a sideband pattern, when the spinning frequency becomes comparable to to strength of the dipolar spin–spin interaction. The theoretical description of this effect is reviewed, including transformation properties of dipolar Hamiltonians under sample spinning conditions and considering the statistical approach based on Mori's formalism and the Anderson-Weiss approximation.

Because the appearance of sharp matching sidebands imposes experimental difficulties in maintaining an efficient polarization transfer, much effort has been invested to modify the original CP scheme and to circumvent the matching sideband problem. The review considers several classes of experiments; (i) mechanical solutions, like off-magic-angle spinning or stop-and-go spinning, (ii) r.f.-based solutions, including phase and amplitude variations of the r.f. fields, and (iii) alternative ways of polarization transfer relying on the nuclear solid effect and dynamic nuclear polarization. The r.f.-based techniques are considered in some more detail, because they do not only allow one to solve experimental problems, but also lead to additional insights into the cross-polarization process in general, and are the basis for some spectral editing tools.

Finally the influence of thermal mobility on cross-polarization is discussed including problems like the achievable rare-spin magnetization as a function of motional correlation times.

# 1 Introduction

## *1.1 Prologue*

The popularity of nuclear magnetic resonance (NMR) for providing details of structure and dynamical processes in molecular systems or solids is largely based on the high-resolution character of most NMR experiments. In NMR studies of liquid samples with high internal mobility this typically means that the dispersion of chemical shifts and the magnitude of indirect spin–spin (J) coupling constants are considerably larger than the natural widths of the resonances. For solid samples, the line widths that are normally attainable with available techniques usually obscure the small J couplings being often so extremely useful in liquid-state NMR. Thus, a mostly applied criterion for high resolution in solid samples is the ability to distinguish resonances with reasonably different chemical shifts.

Therefore, it has been a major technical goal in NMR to develop specific techniques for narrowing the resonance lines in solid-state spectra and to achieve a spectral resolution at least comparable to that of liquids in the sense as characterized above. This goal was addressed successfully in the early 1970s and years later, mainly by introduction of three sophisticated experimental techniques: *magic-angle sample spinning (MAS), high-power dipolar decoupling* and appropriate *multiple pulse sequences*. In addition, the *polarization transfer* or (in the terminology of solid-state NMR spectroscopy) *cross-polarization (CP)* technique was developed to overcome the problem of low sensitivity in the solid-state NMR of nuclides with low natural abundance (e.g. the rare spins $^{13}C$, $^{15}N$, $^{29}Si$). Moreover, in favourable cases the CP technique may provide a means of circumventing the time bottleneck caused by inefficient spin–lattice relaxation of rare spins in many solids. Each of these techniques is designed to eliminate or to reduce particular types of magnetic interactions and may be applied alone or in combination. The combined CP-MAS approach, e.g. for carbon-13 with proton decoupling and in the other combinations has become a popular and routine experiment for investigating solid specimens and is widely used in various branches of physics, chemistry and biochemistry [1]. Cross-polarization techniques have their roots in the experiments performed by several groups among them, in particular, those of E. L. Hahn [2,3], C. P. Slichter [4,5], A. G. Redfield [6] and J. S. Waugh [7]. The phenomenon of rotary saturation discussed by Redfield [6] is a resonant absorption phenomenon in which a spin-locked magnetization could be resonantly affected by an additional applied time-dependent field. This technique, in general, seemed to indicate that a resonant transfer to and from a spin-locked system of spins is possible. Hartmann and Hahn [2] were interested in *indirectly detecting rare spins*, as were Lurie and Slichter [4], and the resonant transfer of magnetization from one spin system to another was the method of choice. A variant of such indirect detection schemes has been demonstrated by Mansfield et al. [8] but it has not been used furthermore.

A *direct detection method* of the rare spin magnetization where the rare spin system is polarized by the abundant spins was proposed by Pines, Gibby and Waugh [7]. Although the indirect method is of higher sensitivity, the direct method has some practical advantages and has become so far the current manner of investigating solid-state NMR spectra of systems of rare spins.

A comprehensive survey of the various cross-polarization techniques has been presented by Mehring [9] in his excellent monograph. Hence, as far as the content of this book [9] is concerned, it cannot be the aim of the present review to treat again in detail the basic principles of double resonance experiments and of the theory of cross-polarization outlined in Ref. [9]. Consequently, in the following only a very brief introduction into the phenomenological approach and the treatment of the cross-polarization NMR spectra and the dynamics of cross-polarization and spin-decoupling will be given as initiated by McArthur, Hahn and Walstedt [10] and confirmed later by Demco, Tegenfeldt and Waugh [11] using the Mori formalism. Independent of the discussion of the fundamentals presented in Ref. [9] and in the famous monograph of Ernst, Bodenhausen and Wokaun [12] a short introduction into double resonance technique seems to be necessary in the present review in order to understand the peculiarities in cross-polarization and related phenomena due to the influence of sample rotation which shall be the aim of this article.

## *1.2 Scope, Organization and Delimitation of the Contribution*

The *scope and the organization* of this contribution are as follows: Cross-polarization in non-rotating solids will be the subject of Sect. 2. In order to introduce adequately into the phenomenon of cross-polarization it seems to be necessary to deal at first with a description of the standard spin-lock cross-polarization transfer (Sect. 2.1) proposed by Pines, Gibby, and Waugh [7] for the direct detection of rare spin NMR spectra in solids where the Zeeman order in the rotating frame of the abundant I spins in transferred to the Zeeman order in the rotating frame of the rare S spins under the conditions of Hartmann-Hahn match. The matching condition requires that the angular frequencies, $\omega_{1I}$ and $\omega_{1S}$ of the coupled *I* and *S* spins in the rotating frame, respectively, are equal, $\omega_{1I} = \omega_{1S}$ or (for off-resonance $\Delta\omega_i \neq 0$) $\omega_{eI} = \omega_{es}$ with $\omega_{ei} = [\omega_{1i}{}^2 + \Delta\omega_i{}^2]^{1/2}$, $i = I, S$. The different steps of the CP experiment are treated as well as the calculation of the cross-polarization transfer rate where as usually at first a simplified thermodynamic model is applied (Sect. 2.2). The cross-polarization will be described as a thermal equilibration between the S spin and the I spin reservoirs under the influence of *I–S* coupling. Hence, a short-time-scale process is considered establishing a thermal equilibrium in each reservoir being characterized by the corresponding spin temperatures. A long-time-scale evolution will lead to a uniform temperature of the two reservoirs and the only constraint is that the system is conservative, i.e. that the energy is conserved is a whole.

The usual approach for the description of the cross-polarization rate is described (Sect. 2.3) which satisfactorily may reproduce the dependence of the experimental time constants on the irradiation strength and the mismatch of the Hartmann-Hahn coupling conditions and which may be applied for the interpretation of selective CP measurements for the study of the local structures in polycrystalline materials. The fact that optimum CP is achieved under Hartmann-Hahn match condition is explained in terms of this treatment only by purely kinetic arguments, i.e. by a rapid decrease of the CP rate away from match so that in fact a quasi-equilibration cannot be achieved. In Sect. 2.4 a modified thermodynamic model is described as proposed by Levitt, Suter, and Ernst [13] which has no parallel in the macroscopic "classical" thermodynamics and which is a consequence of the quantization of the spin systems. This approach leads to a better understanding of the achievable S spin cross-polarization under Hartmann-Hahn mismatch and may be considered as the basis for cross-polarization experiments under mismatched Hartmann-Hahn condition, like MOIST, successive polarization under mismatched Hartmann-Hahn conditions, and cross-polarization experiments using non-Zeeman reservoirs.

Cross-polarization in rotating solids is treated in Sect. 3. First of all it starts with a characterization of effects of sample spinning on cross-polarization (Sect. 3.1). The theoretical description of cross-polarization dynamics in rotating solids involves both the "microscopic" level, i.e. the examination of the modulation (transformation properties) of the dipolar interaction in case of sample spinning, and a statistical description based on a modification of the treatment according to Demco, Tegenfeldt, and Waugh [11]. On this basis the Hartmann-Hahn match can be calculated as a function of the strength of homonuclear and heteronuclear dipolar coupling and of the frequency for magic-angle spinning (MAS). Techniques to solve CP/MAS matching problems are discussed in Sect. 3.3. Different strategies are analysed here to overcome the extreme sensitivity of the CP/MAS spectral intensities with respect to the spinning rate and stability requirements of the amplifier and the probe. The techniques include mechanical methods, r.f.-based techniques and the possibility to use alternative ways of CP transfer. The heterogeneity of cross-polarization is the subject of Sect. 3.4, including both the phenomena in relation to cross-depolarization and polarization inversion as well as more sophisticated techniques of rotor-synchronized amplitude-modulation CP schemes. Finally, in Sect. 3.5 the influence of thermal mobility on the efficiency of the cross-polarization is treated where the validity of the predictions of the "short correlation time theory" is shown also for an extrapolation to the limit of long correlation times, i.e. the limit of solid-state NMR under the condition of which normally CP techniques are considered.

Unavoidably, because of the limited space and the enclosure with other contributions in these volumes about "Solid-State NMR Spectroscopy", there is a lack of several details in this presentation which are in some relation to the phenomena of cross-polarization in combination with and without mechanical sample rotation. Hence, it seems to be worthwhile to discuss the *delimitation*, to mention the points which were omitted here, and to characterize at least for the first item related with quadrupole effects the existing problems and the solutions achieved in a brief general description, because of its considerable practical relevance and its closer relation to Sects. 2 and 3:

- *Cross-Polarization and Mechanical Sample Rotation in the Case of Quadrupolar Perturbed NMR Spectra*

Recently, quadrupole effects in solid-state nuclear magnetic resonance have been extensively reviewed by Freude and Haase [14]. In context with the present review at first the phenomena have to be mentioned (cf. Ref. [14], chapter 6) originating from the application of *techniques of mechanical sample spinning to quadrupolar perturbed NMR spectroscopy.* Technical achievements have to be mentioned, viz. spinning about one axis, like magic-angle spinning [MAS, magic angle of 54.7 °C of the spinning axis with respect to the direction of the external d.c. magnetic field], variable angle spinning [VAS, use of definite variable angles], and dynamic angle spinning [DAS, sample spinning about an axis inclined at a time dependent angle with respect to the direction of the d.c. external magnetic field], and the use of highly sophisticated techniques with a simultaneous rotation about two axes [double-rotation (DOR) NMR spectroscopy, viz. torque-free rotation about an outer axis at the magic angle and an inner axis at 30.55° with respect to the direction of the external d.c. magnetic field in order to eliminate the influence of first and second order quadrupole interactions, respectively, on the NMR line shape]. In context with the present review, in particular the peculiarities in the *excitational phenomena for quadrupolar nuclei* have to be understood (cf., for instance, also chapter II, Sect. 2.2.5 in Ref. [15]) being of great importance for cross-polarization experiments with quadrupolar nuclei. A *nonselective excitation* (measurement of the whole spectrum) can be achieved if the angular frequency $\omega_{1S}$ for quadrupolar nuclei with spin quantum number $S > 1/2$ is much larger than the quadrupole frequency $\omega_Q = 3eq^2Q/[2S(2S-1)\hbar]$, i.e. $\omega_{1S} \gg \omega_Q$, with the usual abbreviations. A *selective excitation* of the transition in resonance $(m \Rightarrow m+1)$ occurs for $\omega_{1S} \ll \omega_Q$. In this case the effective angular frequency (nutation frequency in nutation NMR spectroscopy [14]), $\omega_{1\mathrm{Seff}}$, is increased by the factor

$$\omega_{1\mathrm{Seff}}/\omega_{1S} = [(S-m)(S+m+1)]^{1/2},$$

i.e. by a factor of $S + 1/2$ for the central transition $(m = -1/2)$ of half-integer spins $S$ and a factor of $(2)^{1/2}$ for a single quantum transition $(m \Rightarrow m+1, m = -1)$ in $S = 1$ spin systems, etc. On the basis of the representation of the Hamiltonians and the spin density matrix in terms of single transition operators [or *fictitious spin-1/2 operators*] (see, e.g., [16] to [19]), single quantum (1QT) and double quantum (2QT) cross-polarization in non-rotating samples was treated in previous articles, [20–22]. The theory of spin dynamics for spin $I = 1/2$, $S = 1/2$ systems outlined in the subsequent sections was extended to systems with $I = 1/2$, $S \geqslant 1$. Modified Hartmann-Hahn matching conditions emerged and the dependences of cross-polarization times $T_{IS}$ on r.f. intensity and frequency for spin-locking and adiabatic demagnetization in the rotating-frame experiments were calculated.

For *signle quantum cross-polarization* the Hartmann-Hahn matching condition was found to be different for non-selective and for selective spin-locking of the $S$ spins. For non-selective spin-locking the condition $\omega_{eI} = \omega_{eS}^{1QT}$ holds with the effective angular frequencies, as in standard Hartmann-Hahn cross-polarization, given by $\omega_{eS}^{1QT} = [\omega_{1S}^2 + \Delta\omega_S^2]^{1/2}$. For selective spin-locking cross-polarization, however, instead of $\omega_{1S}$ in the formula for $\omega_{eS}^{1QT}$, the nutation frequency $\omega_{1Seff} = \omega_{1S}[(S-m)(S+m+1)]^{1/2}$ appears, i.e. the Hartmann-Hahn match occurs at

$$\omega_{eS}^{1QT} = [\omega_{1Seff}^2 + \Delta\omega_S^2]^{1/2}.$$

The same theory as described below in Sect. 2 has been applied to compute the cross-polarization rate $1/T_{IS}^{1QT}$ for single quantum cross-polarization of quadrupole nuclei and it was found that $T_{IS}^{1QT} = T_{IS}$ where $T_{IS}$ will be derived subsequently.

For *double quantum cross-polarization* a modified Hartmann-Hahn matching condition,

$$\omega_{eI} = \omega_{eS}^{2QT} = [\omega_{1Seff}^2 + (2\Delta\omega_S)^2]^{1/2},$$

was found containing an effective angular frequency of the $S$ spins being $\omega_{1Seff} = \omega_{1S}^2/\omega_Q$ for $S=1$ systems. Note that the doubled frequency offset appears in the formula for $\omega_{eS}^{2QT}$, so that the mismatch sensitivity, to be later discussed in more detail, is twice that of single quantum transitions. The cross-polarization time $T_{IS}^{1QT} = (1/4)T_{IS}^{2QT}$ is reduced by a factor of 4 compared with single quantum cross-polarization.

General expressions for multi-quantum cross-polarization on spin systems with $I = 1/2$ and $S = 3/2$ in solids (singles-, double-, and triple-quantum transitions) were derived by Vega [22].

The polarization transfer from protons to quadrupole nuclei with half-integer spin quantum number combined with MAS has been demonstrated in several publications, like e.g. $^{11}$B CP/MAS NMR [23], $^{17}$O CP/MAS NMR [24, 25], $^{23}$Na CP/MAS NMR [26], $^{27}$Al CP/MAS NMR [27–33], $^{43}$Ca CP/MAS NMR [34], and $^{95}$Mo CP/MAS NMR [35, 36]. But while the theoretical aspects of $^1$H/Na$^{23}$ cross-polarization transfer in a static single crystal were investigated by Vega [22] and experimental questions of the CP dynamics of powders in some static and MAS experiments were characterized in considerable details as mentioned before, a more detailed theoretical discussion of the r.f.-matching conditions and efficiency of CP for half-integer spins with $S > 1$ in rotating samples has so far, to our knowledge, reported only by Vega [37, 38].

Recently, a polarization transfer from one quadrupole nucleus ($^{27}$Al) to another ($^{17}$O) was reported by Haase et al. [39, 40].

In the remaining part of this section the subjects which will not be considered further will only be mentioned:

- *Two dimensional NMR in rotating solids*
  (see, e.g. [41–74])
- *Spin diffusion without direct relation to the performance of CP/MAS NMR experiments*
  in general and in rotating solids (see, e.g., [75–96])
- *Spinning-sideband manipulation*
  general aspects of modification and suppression of sidebands (see, e.g., [97–113]),
  spinning-sideband calculation and determination of tensor anisotropies (see, e.g., [114–117])
  rotary resonance recoupling and rotational resonance (see, e.g. [118–129])
- *Relaxation in rotating solids*
  (see, e.g. [123–129]).

## 2 Cross-Polarization in Non-Rotating Solids

In the following considerations, only systems are discussed in which two different spin species *I* and *S* are present and where there is the possibility of dipolar and scalar couplings between them. This heteronuclear dipole-dipole coupling may obscure weak interactions that are useful for the characterization and analysis of the structure of the systems investigated by NMR techniques. Such a situation will very often be the case in the investigation of NMR spectra of spins in low natural abundance ("rare" or "dilute" spins *S*) whose dominant interaction is via dipolar couplings to a system of abundant spins *I*. As is well known, to obtain the information inherent in these interaction mechanisms, the selective removal or alteration of the broadening mechanisms can be attained both by manipulation of spin variables and by affecting the spatial variables. Cross-Polarization in the rotating frame is the usual technique today for enhancing the nuclear spin polarization of rare spins and/or isotopes of low magnetogyric ratio, such as $^{13}C$, $^{15}N$, and $^{29}Si$ in solids. There are two common experimental procedures for cross-polarization between spin species *I* and *S*. They involve either adiabatic demagnetization in the rotating frame (ADRF) of the abundant spin reservoir or simultaneous spin-locking of the two species as first proposed by Hartmann and Hahn. Especially in combination with heteronuclear decoupling, the latter technique is experimentally simpler and this direct detection technique is in fact the only practical method for a simultaneously applied magic-angle spinning (MAS) of the sample. These advantages enabled truly routine high-resolution solid-state NMR spectroscopy. For this review it is taken into account that because of these reasons in the overwhelming majority of applications the direct detection of rare spin NMR is preferred. Thus, only the spin-lock polarization transfer technique proposed by Pines, Gibby and Waugh (henceforth abbreviated as the "PGW-technique") is considered, in combination with high-power dipolar decoupling (to manipulate the spin variables) and with mechanical spinning of the sample about axes inclined at a distinct angle with respect to the direction of the magnetic field $B_0$ (MAS or off-MAS) in order to affect the spatial variables of the coupling Hamiltonians.

In the following introduction, the common theoretical description of the spin-lock cross-polarization is outlined first. This is the description found in many publications and textbooks (cf., e.g., [130] and [131]). This theory describes the systems by separate $I$ and $S$ spin reservoirs which can exchange energy in arbitraily small units. The main features of this theory are the attainment of a thermal equilibrium state between the reservoirs under the influence of the $I$–$S$ coupling and the assumption that the total energy is the only relevant constant of the motion. This is equivalent to a maximization of entropy under the single constraint that the total energy remains constant.

The common theoretical description refers for details to Refs. [130] and [131] and starts with an explanation of the PGW-technique, i.e.

- a short explanation of the different steps for the practical realization of the spin-lock polarization transfer (Sect. 2.1),
- a brief discussion of the phenomenological approach of cross-polarization dynamics (Sect. 2.2),
- a review of the theory of cross-polarization rate without taking into account, at first, the influence of magic-angle sample spinning or of internal thermal mobility (Sect. 2.3).

Following this line, the interested reader may easily understand that the simple thermodynamic model described is essentially only applicable exactly at the Hartmann-Hahn match and predictions made using it away from the Hartmann-Hahn match are incorrect. The modified spin thermodynamic theory capable of treating cross-polarization under mismatch Hartmann-Hahn conditions, will be described in Sect. 2.4 closely following a paper of Levitt, Suter and Ernst [133].

## 2.1 The Spin-Lock Polarization Transfer

### 2.1.1 Description of the Spin Systems

We consider the thermodynamics and kinetics of polarization transfer between an ensemble of $N_I$ spins, with magnetogyric ratio $\gamma_I$, and an ensemble of $N_s$ spins, with magnetogyric ratio $\gamma_S$ in a way which is close to Refs. [130, 131, 133]. Since the $I$ spins are assumed to be abundant and the $S$ spins are rare ($N_S \ll N_I$), $S$–$S$ interactions may be ignored and it is valid to treat the system as an ensemble of subsystems each consisting of a single $S$ spin and a large number of $I$ spins. We separate the Hamiltonian of the spin system into a set of thermodynamic reservoir terms denoted by $H_i^0$ and a smaller perturbation term $H^1$,

$$H = H^0 + H^1 \qquad H^0 = \sum_i H_i^0 \tag{1}$$

where the different $H_i^0$ commute

$$[H_i^0, H_j^0] = 0 \tag{2}$$

but

$$[H^0, H^1] \neq 0.$$

The Hamiltonian $H$ is considered to include the principal contributions to the energy of the spin system and it is assumed that $H^1$ plays a purely kinetic role for the transfer of the spin order between the thermodynamic reservoirs $H_i^0$. These reservoir terms are usually separated such that

$$\mathrm{tr}\{H_i^{0+} H_j^0\} \equiv (H_i^0 | H_j^0) = 0, \quad i \neq j. \tag{3}$$

In the "double rotating" frame [130] which rotates at frequencies $\omega_I$ and $\omega_S$ close to the Larmor frequencies $\omega_{0I}$ and $\omega_{0S}$, respectively, the Hamiltonian $H$ is given by

$$H = H_I^0 + H_S^0 + H_{II}^0 + H_{IS}, \tag{4}$$

where

$$\begin{aligned} H_I^0 &= \sum_k \Omega_k I_{kz} + \omega_{1I} \sum_k I_{kx} \\ H_S^0 &= \Omega_S S_z + \omega_{1S} S_x \\ H_{II}^0 &= \sum_j \sum_k d_{jk} [2 I_{jz} I_{kz} - \tfrac{1}{2}(I_j^+ I_k^- + I_j^- I_k^+)] \\ H^1 &= H_{IS}^0 = \sum_k b_k 2 I_{kz} S_z. \end{aligned} \tag{5}$$

As usual, non-secular parts of $H_{IS}$ and $H_{II}$ were omitted. The parameters introduced have the usual meanings: $\Omega_k$ and $\Omega_S$ are resonance offsets, $\omega_{1I} = \gamma_I B_{1I}$ and $\omega_{1S} = \gamma_S B_{1S}$ are related to the radio frequency (r.f.) field amplitudes $B_{1I}$ and $B_{1S}$. The heteronuclear dipolar coupling constants $b_k$ between the spins $S$ and $I_k$ (in units of $s^{-1}$), and the homonuclear dipolar coupling constants $d_{jk}$, are given by

$$\begin{aligned} b_k &= -\frac{\mu_0 \gamma_I \gamma_S \hbar^2}{4\pi r_k^3} \cdot \frac{1}{2}(3\cos^2\theta_k - 1) \\ d_{jk} &= -\frac{\mu_0 \gamma_I^2 \hbar^2}{4\pi r_{jk}^3} \cdot \frac{1}{2}(3\cos^2\theta_{jk} - 1). \end{aligned} \tag{6}$$

where $r_k$, and $r_{jk}$ are internuclear distances and $\theta_k$, $\theta_{jk}$ are angles between internuclear vectors and the d.c. field $B_0$. Scalar couplings have been omitted.

In the case of the strong r.f. field such that $|\omega_{1I}| \gg |d_{jk}|$ and $|\omega_{1S}| \gg |b_k|$ it is convenient to represent the Hamiltonian in a "tilted" frame rotated around the $y$ axis by an angle $\pi/2$, i.e. in frames whose $z$ axes are parallel to the instantaneous direction of the rotating r.f. fields. An arbitrary operator $A^T$ in this frame is defined by

$$A^T = R^{-1} A R \tag{7}$$

where $A$ is respective operator in the "doubly rotating" frame and

$$R = \exp\left[-i\frac{\pi}{2}\left(\sum_k I_{ky} + S_y\right)\right] \tag{8}$$

Thus we arrive at the Hamiltonian in the tilted frame

$$H^T = H_I^T + H_S^T + H_{IS}^T + H_{II}^T, \tag{9}$$

with the operators

$$\begin{aligned} H_I^T &= \sum_k \omega_{1I} I_{kz} \\ H_S^T &= \omega_{1S} S_z \\ H_{IS}^T &= \sum_k b_k 2I_{kx} S_x \\ H_{II}^T &= H_{II,sec}^T + H_{II,nonsec}^T \end{aligned} \tag{10}$$

and

$$\begin{aligned} H_{II,sec}^T &= -\frac{1}{2}\sum_{j,k} d_{jk}[2I_{jz}I_{kz} - \tfrac{1}{2}(I_j^+ I_k^- + I_j^- I_k^+)] \\ H_{II,nonsec}^T &= \frac{3}{2}\sum_{j,k} d_{jk}\tfrac{1}{2}(I_j^+ I_k^+ + I_j^- I_k^-) \end{aligned} \tag{11}$$

where $d_{jk}$ and $b_k$ are given by Eq. (6). $H_{II,sec}^T$ and $H_{II,nonsec}^T$ are the parts of homonuclear dipolar interaction Hamiltonian $H_{II}^T$ being secular and nonsecular with respect to the Hamiltonian $H_I^T$, respectively. For large r.f. fields $B_{1I}$ with $|\omega_{1I}| = |\gamma_I B_{1I}| \gg |d_{jk}|$ the nonsecular Hamiltonian $H_{II,nonsec}^T$ may be dropped. The reservoir terms according to Eq. (2) are then $H_i^0 = H_I^T, H_S^T, H_{II,sec}^T$ and the term $H^1$ is given by $H_{IS}^T$.

A somewhat more complicated situation may arise if the "tilted rotating" frame is defined to be aligned along the directions of the effective fields in the rotating frame inclined at angles $\theta_I$ and $\theta_S$ with respect to the d.c. field $B_0$. Instead of Eq. (8) the transformation is given by

$$R = \exp\left[-i\theta_I \sum_k I_{ky}\right] \exp[-i\theta_S S_y] \tag{12}$$

where

$$\tan\theta_j = \frac{\omega_{1j}}{\Delta\omega_j}, \quad j = I, S \tag{13}$$

$$\Delta\omega_j = \omega_{0j} - \omega_j, \quad \omega_{ej} = \sqrt{\omega_{1j}^2 + \Delta\omega_j^2}. \tag{14}$$

The respective terms are

$$\begin{aligned} H_I &= -\omega_{eI} I_z \\ H_S &= -\omega_{eS} S_z \end{aligned}$$

$$H_{IS} = \cos\theta_I \cos\theta_S H_{IzSz} + \sin\theta_I \sin\theta_S \cdot H_{IxSx} - \sin\theta_I \cos\theta_S H_{IxSx} - \cos\theta_I \sin\theta_S H_{IzSx} \tag{15}$$

$$H_{II} = P_2(\cos\theta_I) H_{II}^0 + H_{II,\mathrm{nonsec}}$$

with

$$P_2(\cos\theta_I) = \tfrac{1}{3}(3\cos^2\theta_I - 1), \quad H_{I_\alpha S_\beta} = \sum_k b_k 2 I_{\alpha k} S_\beta, \quad (\alpha, \beta) = (x, y), \tag{16}$$

and $H_{II}^0$ is given in Eq. (5) and $b_k$ is chosen as in Eq. (6). For the same reason as discussed already, the nonsecular term of $H_{II}$ may be dropped.

### 2.1.2 The Thermodynamic Model

To describe the resonant transfer of magnetization between the spin systems it is useful to consider the magnetization $M_i (i = x, y, z)$, the energy $E$, and the entropy $\zeta$ which are given by the expressions

$$M_{Ii} = \gamma\hbar \,\mathrm{Tr}\{\varrho I_i\} \quad \text{(e.g. for } I \text{ spins)} \tag{17}$$

$$E^j = \hbar \,\mathrm{Tr}\{\varrho H_j\} \quad (j = I, S) \tag{18}$$

$$\zeta = -k \,\mathrm{Tr}\{\varrho \ln \varrho\}. \tag{19}$$

The operation Tr denotes the appropriate traces. $\varrho$ is the spin density matrix operator, $I_i$ and $S_i$ are dimensionless spin operators, and $H$ is the Hamiltonian (the eigenvalues of which are taken in units $s^{-1}$ of angular frequencies). In the high-temperature limit usually realized in the experiments the density matrix is given by

$$\varrho = 1 - \frac{\hbar H}{kT} \tag{20}$$

and for spin systems subject to the Zeeman interaction with $B_0$ along the $z$ axis of the laboratory frame the quantities become $(j = I, S)$

$$M_{0j} = \frac{C_j B_0}{T} \tag{21}$$

$$E^j = \frac{C_j B_0^2}{T} \tag{22}$$

$$\zeta^j = \frac{C_j B_0^2}{T^2} + \mathrm{const} \tag{23}$$

with the Curie constant

$$C_I = \frac{\gamma_I^2 \hbar^2 I(I+1) N_I}{3k}$$

and an analoguous expression for $C_S$. For purposes of discussing polarization

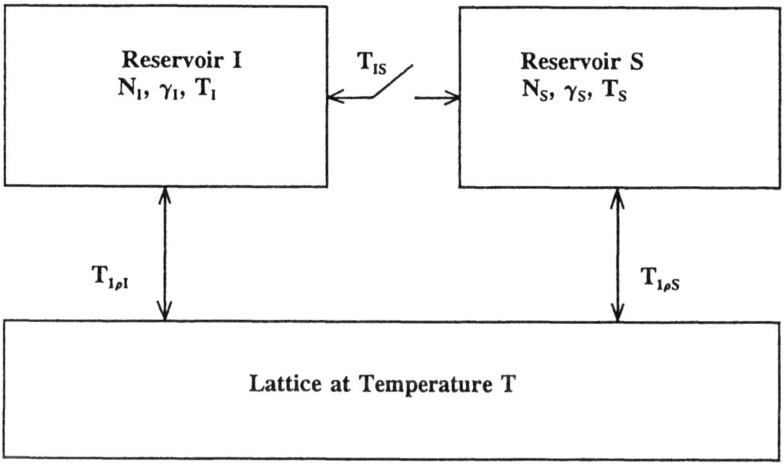


**Fig. 1.** Spin thermodynamics of the spin-lock cross-polarization transfer. In this simple approach the spin systems form reservoirs that can be in contact with each other (cross-polarization rate $T_{IS}^{-1}$) and with the lattice temperature $T$. $T_{1\rho I}$ and $T_{1\rho S}$ are the relaxation rates in the rotating frame

transfer in terms of spin thermodynamics the magnetizations of the subsystems are characterized by the spin temperatures $T_I$ and $T_S$, $M_{I0} = C_I B_0 / T_I$ and $M_{S0} = C_s B_0 / T_S$. The $I$ and $S$ spin systems form reservoirs (at temperature $T_I$ and $T_S$, respectively) that can be in contact with each other and with the lattice (L) which is suggested to have an infinitely high heat capacity (at temperature $T$) (Fig. 1). With respect to the concept of spin temperature in nuclear magnetic resonance we refer to Goldman's excellent monograph [132]. The rate of polarization transfer between the $I$ and $S$ reservoirs is characterized by a time constant $T_{IS}$. Under conditions of cross-polarization $T_{IS}$ denotes the cross relaxation time and the inverse value $T_{IS}^{-1}$ is the cross-polarization (transfer) rate. In the presence of the r.f. fields $B_{1I}$ and $B_{1S}$ (spin-lock conditions) the contact between the spin reservoirs and the lattice is characterized by the spin-lattice relaxation times $T_{1\varrho I}$ and $T_{1\varrho S}$ in the rotating frame.

### 2.1.3 The Different Steps

The *following steps* in a polarization transfer experiment can be specified (Fig. 2):

(a) reduction of the spin temperature of the $I$-spin reservoir
(b) transfer of polarization between the $I$- and $S$-spin reservoirs and
(c) observation of the $S$-spin magnetization.

ad (a):
Assume that the spin system $I$ is in thermal equilibrium with the lattice ($T_I = T$) before the first radio frequency (r.f.) irradiation. Immediately after the $90^\circ_x$ pulse

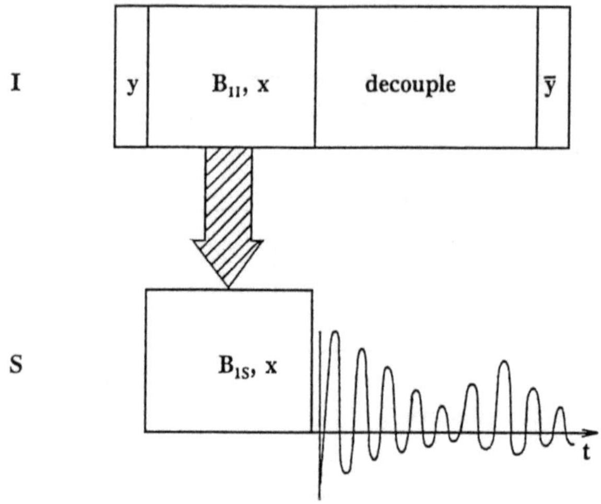


**Fig. 2.** Schematic representation of the pulse sequence for a spin-lock cross-polarization experiment. The spins $I$ are abundant and the spins $S$ are rare, $N_I \gg N_S$

is applied to the spin system $I$ and at the instant when the $I$ spins begin to be locked in the field $B_{1I}$ along the $y$ direction in the rotating frame, these spins have a low spin temperature $T_{0I}$: We assume that the $90^\circ_x$ pulse is so rapid that the change in the $I$-spin reservoir is adiabatic and reversible. Then a conservation of the entropy of the $I$ spin system can be suggested which implies

$$\frac{B_0}{T} = \frac{B_{1I}}{T_{0I}}. \tag{24}$$

Since usually the amplitude $B_{1I}$ of the r.f. field is much less than the induction $B_0$ of the d.c. field, at the beginning of the spin-lock period the $I$ spin temperature $T_{0I}$ is given by $T_{0I} = (B_{1I}/B_0)\, T$, i.e. $T_{0I}$ is very low relative to the lattice temperature $T$.

ad (b):
In presence of the r.f. fields with amplitudes $B_{1I}$ and $B_{1S}$ applied at the Larmor frequencies $\omega_I$ and $\omega_S$, respectively, the polarization transfer takes places at a rate $T_{IS}^{-1}$ which will be discussed below. If the relaxation rates $T_{1\varrho I(S)}^{-1}$ in the rotating frame are slow compared to the rate $T_{IS}^{-1}$, the coupled $I$–$S$ spin systems can be considered to be isolated from the lattice. Thus, the polarization transfer is only related with heating the $I$- and cooling the $S$-spin system, i.e. energy conservation. If the equilibrium temperature of the coupled $I$–$S$ spin system is denoted by $T_{eq}$, this requires

$$E^I(T_{0I}) - E^I(T_{eq}) = E^S(T_{eq}) - E^S(T) \tag{25}$$

or, taking Eq. (22) for $E^I$ and $E^S$ with $B_{1I}$ and $B_{1S}$, respectively, instead of $B_0$,

$$C_I B_{1I}^2 \left( \frac{1}{T_{0I}} - \frac{1}{T_{eq}} \right) = C_S B_{1S}^2 \left( \frac{1}{T_{eq}} - \frac{1}{T} \right), \tag{26}$$

or

$$T_{eq} = T_{0I} \frac{1+\varepsilon}{1+\frac{T_{0I}}{T}\varepsilon} \tag{27}$$

with

$$\varepsilon = \frac{C_S B_{1S}^2}{C_I B_{1I}^2} = \frac{N_S \gamma_S^2 B_{1S}^2 S(S+1)}{N_I \gamma_I^2 B_{1I}^2 I(I+1)}. \tag{28}$$

Since $T_{0I} \ll T$ and $N_S \ll N_I$ holds ($S$ is a by definition a rare spin!) we arive at

$$T_{eq} \cong (1+\varepsilon)T_{0I} = (1+\varepsilon)TB_{1I}/B_0. \tag{29}$$

ad (c):
To consider the *kinetics of polarization transfer* it will be shown that an efficient cross-polarization is achieved if the *Hartmann-Hahn match condition*

$$\gamma_S B_{1S} = \gamma_I B_{1I} \tag{30}$$

holds. Normally it is assumed (see Sect. 2.3) that the rate of cross-polarization decreases rapidly away from Hartmann-Hahn match so that in fact the equilibrium is not attained except when the r.f. nutation frequencies are equal, Eq. (30). Independent of these kinetic arguments, unless Eq. (30) is valid, the spacing between the energy levels of $H_I$ and $H_S$ (neglecting resonance offsets) does not match. Hence it is obviously impossible to allow exchange of single quanta between these reservoir unless Eq. (30) holds if simultaneously energy conservation is taken into account. This is to be considered the crucial point of the simple thermodynamic model. The enhancement $M_S/M_{0S}$ of the signal-to-noise ratio in this CP NMR experiment ($\sim M_S$) relative to a simple measurement ($\sim M_{0S}$) of a free induction decay (FID) associated with the cooling of the $S$-spin reservoir to $T_{eq}$ directly results from the Curie law [131]: At the spin temperature $T_{eq}$ the magnetization $M_S$ of the $S$-spin system in a field $B_{1S}$ is (using Eqs. (28) and (29) and taking $\varepsilon \ll 1$)

$$M_S = \frac{C_S B_{1S}}{T_{eq}} = \frac{C_S B_0}{T(1+\varepsilon)} \frac{\gamma_I}{\gamma_S} \cong M_{0S} \frac{\gamma_I}{\gamma_S}. \tag{31}$$

Hence, the gain $G$

$$G = \frac{M_S}{M_{0S}} \cong \frac{\gamma_I}{\gamma_S}, \tag{32}$$

is comparable with the nuclear Overhauser effect (NOE) enhancement factor in case of the dynamic polarization of spins $S$ in a conventional NOE FT NMR experiment for liquids under ideal conditions. For a more precise approach the conditions of the CP NMR experiment must be specified. If during a single spin-lock experiment $n$ contacts can be realized, a maximum value (see Ref. [130], chapter 4) may be achieved under the condition $n\varepsilon \approx 1.3$. For organic solids with natural abundance of $^{13}C$ nuclei the ratio of proton and carbon nuclei is $N_I/N_S \approx 150$ and the factor

$$G = 0.64 \frac{\gamma_I}{\gamma_S} \sqrt{\frac{1}{\varepsilon}} \tag{33}$$

may be as large as $G \cong 30$. Since in most investigations, however, single-contact CP experiments are performed, the multi-contact experiment will not be considered further here.

## 2.2 *The Phenomenological Approach for the Description of the CP Dynamics*

The approach relies on the following assumptions:

- We assume as before a coupled $I$–$S$ spin system with relaxation rates $T^{-1}_{1\rho I(S)}$ in the rotating frame which are slow with respect to the cross-polarization rate $T^{-1}_{IS}$, i.e. we consider at first an $I$–$S$ spin system isolated from the lattice for which the total spin energy is conserved.
- We neglect oscillations of the nuclear magnetization of spins $S$ for short contact times. Such oscillations can occur in the case of discrete $I$–$S$ dipolar spectra.
- We consider only an experimental situation where MAS and off-MAS technique are not applied or do not interfere the CP process.

Under these conditions and assuming first order kinetics, the changes of the inverse spin temperatures

$$\beta_i(t) = \hbar/(kT_i(t)) \tag{34}$$

being proportional to the magnetization $M_i(t)$ of the spin systems $i = I, S$ can be described by the equation

$$\begin{aligned} \dot{\beta}_S &= -\frac{1}{T_{IS}}(\beta_S - \beta_I) \\ \dot{\beta}_I &= -\frac{\varepsilon}{T_{IS}}(\beta_I - \beta_S) \end{aligned} \tag{35}$$

where $\varepsilon$ is defined as given above.

The influence of rotating frame relaxation usually is included by adding the relaxation terms $-\beta_i/T_{1\rho i}(i = I, S)$ leading to the coupled equations

$$\begin{aligned} \dot{\beta}_S &= -\frac{1}{T_{IS}}(\beta_S - \beta_I) - \frac{1}{T_{1\rho S}}\beta_S \\ \dot{\beta}_I &= -\frac{1}{T_{IS}}(\beta_I - \beta_S) - \frac{1}{T_{1\rho I}}\beta_I . \end{aligned} \tag{36}$$

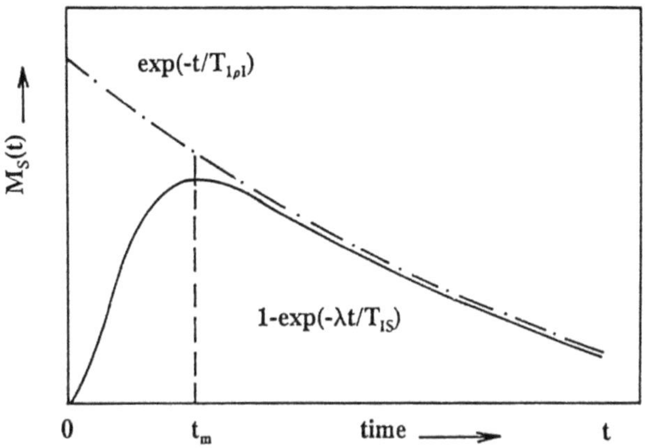


**Fig. 3.** Typical representation of the $S$ spin magnetization in a spin-lock cross-polarization experiment. The $S$ spin magnetization reaches a maximum at time $t_m$ (see text)

In agreement with the preposition of the common (simple) thermodynamic model we consider only the situation when the Hartmann-Hahn condition is fullfilled and when an extremely diluted spin system $S$ is present ($\varepsilon = 0$). Since for a usual CP NMR experiment described above the initial conditions are $\beta_I(0) = \beta_{I0}$ and $\beta_S(0) = 0$, the spin systems evolve according to

$$\begin{aligned} \beta_I(t) &= \beta_{I0} \exp(-t/T_{1\rho I}) \\ \beta_S(t) &= (\beta_{I0}/\lambda) \exp(-t/T_{1\rho I})[1 - \exp(-\lambda t/T_{IS})] \end{aligned} \tag{37}$$

where

$$\lambda = 1 + \frac{T_{IS}}{T_{1\rho S}} - \frac{T_{IS}}{T_{1\rho I}}. \tag{38}$$

The typical behaviour of the S spin magnetization is shown in Fig. 3.

The time evolution of the S spin magnetization is dominated by cross-polarization for short transfer times $t$. In the case of negligible $I$ spin relaxation in the rotating frame ($T_{IS}/T_{1\rho I} \ll 1$) and under the condition of Hartmann-Hahn match, the asymptotic value is given by

$$M_S(t = \infty)/M_{S0} \cong \frac{\gamma_I}{\gamma_S}. \tag{39}$$

Suggested that $T_{IS} < T_{1\rho I} \ll T_{1\rho S}$ holds the S spin magnetization reaches a maximum at time $t_m$,

$$t_m = \frac{T_{IS} \cdot T_{1\rho I}}{T_{1\rho I} - T_{IS}} \ln \frac{T_{1\rho I}}{T_{IS}}. \tag{40}$$

For larger transfer times the time evolution is only governed by the $I$ spin relaxation in the rotating frame (Fig. 3).

One of the problems of using CP NMR techniques to study spin systems with low magnetogyric ratios and low natural abundance is *quantitation.* Because the CP rates $T_{IS}^{-1}$ depend on the nature of the interactions, they may differ for various nuclei in the same molecule. Quantitative measurements must, therefore, include some knowledge of the rate $T_{IS}^{-1}$ of the various spins. In the absence of a knowledge of these rates, one has to check the consistency of intensities measured under the condition of a spin-lock CP NMR experiment. This problem is of particular importance in systems with heterogeneous distribution of the abundant spins, e.g. protons in many amorphous materials such as coal, amorphous carbon layers, amorphous silicon and glassy silicon dioxide. Protons in such materials are used as sources of nuclear polarization for rare spins ($^{13}C$ and $^{29}Si$). The following possibilities are frequently used to solve the question of whether all rare spins are being consistently observed, viz.

(i) by comparing the results from CP and ordinary FID NMR spectra,
(ii) by using a multiple cross-polarization (MCP) technique.

ad (i):

This way is rather time consuming. Furthermore, given the possibility of a broad distribution of relaxation times for the rare spins, FID spectra would have to be obtained for a series of relaxation delay times until there are no variations in the relative peak intensities. Furthermore, reliable CP NMR measurements require a broad variation of CP contact times in order to adequately measure proton-poor and proton-rich regions with longer and shorter time constants $T_{IS}$, respectively. Contact times of several hundreds of miliseconds, however, are associated with experimental difficulties and more significantly, short proton $T_{1\rho I}$ values could militate against such experiment.

ad (ii):

The utility of a multiple CP technique was first discussed by Melchior et al. [134] and further examined in Ref. [135]. The pulse scheme is shown in Fig. 4. A chain of N conventional cross-polarization steps is applied each of which consists (in the $I$ channel) of a $90^\circ_x$ pulse, a spin-lock-pulse (along the $y$ axis of the rotating frame), and a $90^\circ_{-x}$ pulse to restore the abundant $I$ spin magnetization along the $z$ axis. In the $S$ channel each spin-lock pulse is also followed by a $90^\circ_{-x}$ pulse to restore the $S$ spin magnetization along the $z$ axis. The steps are separated by intervals of about five times the spin $I$ spin relaxation time $T_1$ to enable thermal equilibration for the abundant spins. The spin magnetization grows steadily during the process and is observed at the end of the sequence in the common way. The time dependence of the inverse spin temperature has the standard solution of the form [135]

$$\beta_S(t) = \beta_+ e^{-\lambda_+ t} + \beta_- e^{-\lambda_- t}, \tag{41}$$

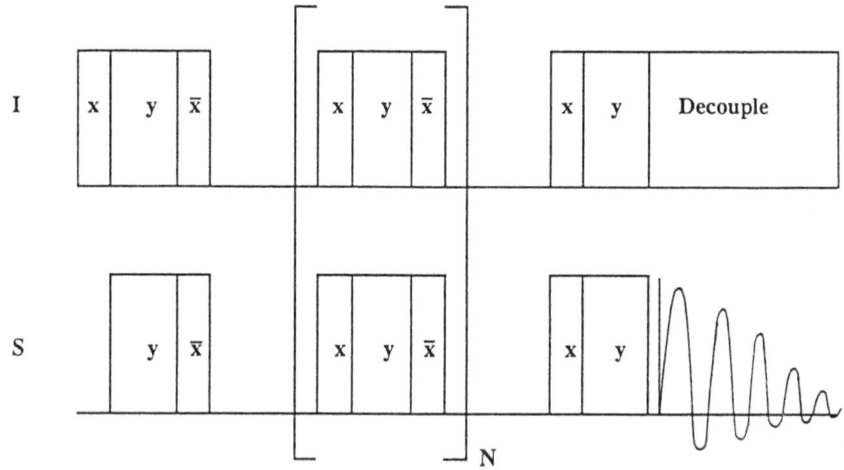


**Fig. 4.** Multiple cross-polarization technique for the study of heterogeneous materials. The sequence consists of a chain of conventional cross-polarization steps separated by intervals of the order of $5T_{11}$ during which the $S$ spin magnetization is kept along the z axis

where

$$\lambda_{\pm} = \tfrac{1}{2}A \pm \tfrac{1}{2}[A^2 - 4B]^{1/2}$$
$$\beta_{\pm} = \frac{\pm(1/T_{\mathrm{IS}})(\beta_{\mathrm{SO}} - \beta_{\mathrm{IO}}) - \beta_{\mathrm{SO}}\lambda_{\pm}}{(\beta_{+} - \beta_{-})}. \tag{42}$$

$\beta_{\mathrm{IO}}$ and $\beta_{\mathrm{SO}}$ are the initial inverse spin temperatures (for each step) and the rate

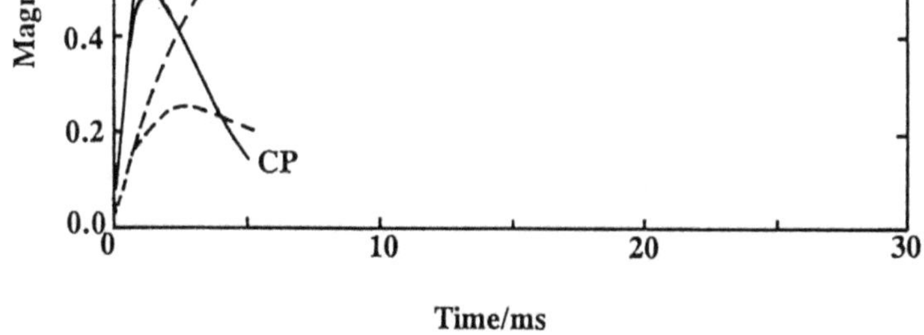


**Fig. 5.** Comparison of normal single contact cross-polarization (CP) and multiple cross-polarization (MCP) experiments. In the calculations different values for $T_{\mathrm{IS}}$ were used, $T_{\mathrm{IS}} = 1$ ms (*solid line*) and 4 ms (*dashed line*) [135]

constants are given by

$$A = \frac{1}{T_{IS}} + \frac{1}{T_{1\rho I}}, \tag{43}$$

$$B = \frac{1}{T_{IS} T_{1\rho I}}. \tag{44}$$

A negligible $S$ spin relaxation rate ($T_{1\rho S}^{-1} = 0$) is always assumed. A schematic time evolution of the $S$ spin magnetization (as function of the aggregate contact time, $t = N t_{CP}$) is shown in Fig. 5. Further polarization schemes are discussed in Sect. 2.4.

## 2.3 The Cross-Polarization Rate

A suitable description of cross-polarization transfer rate $T_{IS}^{-1}$ is necessary in order to account for the fact that $T_{IS}^{-1}$ decreases monotonically to zero for increasing

$$\begin{aligned} \Delta\omega &= \omega_{1S} - \omega_{1I}, \\ \omega_{1j} &= \gamma_j B_{1j} \qquad (j = I, S), \end{aligned} \tag{45}$$

i.e. with increasing mismatch of the Hartmann-Hahn condition.

As mentioned before, $B_{1S}$ and $B_{1I}$ are the amplitudes of the r.f. fields in the rotating frame of the $I$ and $S$ spins, respectively. Instead of $\omega_{1j}$ also effective frequencies,

$$\omega_{ej} = (\omega_{1j}^2 + \Delta\omega_j^2)^{1/2} \tag{46}$$

with the resonance offset $\Delta\omega$, may occur in the Hartmann-Hahn condition. Most of importance for the following discussion is that the transfer rate $T_{IS}^{-1}$ is proportional to a spectral distribution function $J(\Delta\omega_e)$ for the cross-polarization process

$$T_{IS}^{-1} \sim J(\Delta\omega_e) \tag{47}$$

which, under spin-lock (SL) conditions, in most cases can be assumed to be proportional to a Gaussian shaped function [130]

$$T_{IS}^{-1} \sim J_{SL}(\Delta\omega_e) \sim \exp(-\tfrac{1}{4}\Delta\omega_e^2 \tau_c^2) \tag{48}$$

with a correlation time $\tau_c$ yet to be specified. Note that under conditions of adiabatic demagnetization in the rotating frame (ADRF), the experimental transfer rate mostly depends exponentially on $\omega_{es}$ [130]:

$$T_{CP}^{-1}(\mathrm{ADRF}) \sim J_{\mathrm{ADRF}}(\omega_{es}) \sim \exp(-\omega_{es}\tau_c) \tag{49}$$

This experimental behavior was first discussed by McArthur et al. [136]. The theoretical treatment described in the following has been supplied by Demco, Tegenfeldt and Waugh [137] using the memory function approach by Mori and Zwanzig [138, 139]. Instead of the phenomenological equations (35) the treatment starts from the Mori type integro-differential equations [130]

$$\frac{d\beta_i(t)}{dt} = -\left\{\int_0^t dt' K_{ii}(t-t')\beta_i(t') - \int_0^t dt' K_{ij}(t-t')\beta_j(t')\right\} \tag{50}$$

$i \neq j,\ i, j = I, S,$
where the kernels $K_{ii}(t-t')$ and $K_{ij}(t-t')$ represent the "memory" effects taking into account the history of the CP process at earlier times. For this reason $K_{ii}$ and $K_{ij}$ are denoted as "memory function".

The usual approach suggests the fact correlation limit $\tau_c \ll t,\ t-t' = \tau$ to be valid for which in the integrals $\beta_i(t') \approx \beta_i(t)$ may be replaced (i.e. neglect the time retardation) and for which the upper boundary may be replaced by $t \to \infty$.

Thus, from Eq. (50) we may construct the same type of differential equations as given in Eq. (35),

$$\frac{d\beta_i(t)}{dt} = -\beta_i(t)\int_0^\infty d\tau\, K_{ii}(\tau) + \beta_j(t)\int_0^\infty d\tau\, K_{ij}(\tau). \tag{51}$$

Taking, for example, the rate equation for $i = S,\ j = I$ and comparing Eqs. (51) and (35), the following expression for the cross-polarization rate can be obtained [130]:

$$\frac{1}{T_{IS}} = \int_0^\infty d\tau\, K_{SS}(\tau). \tag{52}$$

In case of spin-lock CP treated here, the memory function $K_{SS}(\tau)$ is given by [130]

$$\begin{aligned} K_{SS}(\tau) = {} & \sin^2\theta_I \sin^2\theta_S \cdot M_2^{IS} \cdot C_x(\tau) \cdot \tfrac{1}{2}[\cos(\omega_{eS}-\omega_{eI})\tau + \cos(\omega_{eS}+\omega_{eI})\tau] \\ & + \cos^2\theta_I \cdot \sin^2\theta_S \cdot M_2^{IS} \cdot C_z(\tau) \cdot \cos(\omega_{eS}\tau) \end{aligned} \tag{53}$$

with the usual abbreviations

$$\tan\theta_I = \frac{\omega_{1i}}{\Delta\omega_i},$$

the resonance offset $\Delta\omega_i = \omega_{0i} - \omega_i$, and the effective angular frequencies $\omega_{ei} = \sqrt{\omega_i^2 + \omega_{1i}^2}$ in the rotating frame $(i = I, S)$. $M_2^{IS}$ is the Van Vleck second moment of the $I$–$S$ dipolar coupling. Usually Eq. (53) is simplified to

$$K_{SS}(\tau) = \tfrac{1}{2}\sin^2\theta_I \sin^2\theta_S \cdot M_2^{IS} \cdot C_x(\tau) \cdot \cos(\omega_{eS} - \omega_{eI})\tau \tag{54}$$

since the respective cross-polarization spectral densities

$$J_i(\omega) = \int_0^\infty d\tau \cdot \cos\omega\tau \cdot C_i(\tau) \tag{55}$$

approach zero if $\omega \gg \omega_{loc,I} \simeq \tau_c^{-1}$. This is valid for the second and third term in Eq. (53) because the corresponding frequencies $\omega = \omega_{eS} + \omega_{eI}$ and $\omega = \omega_{eS}$ are much larger than the frequencies $\omega_{loc,I}$ being proportional to the strength of the local $I$–$I$ dipolar interactions. Mostly of importance for the discussion in the next chapters is the result that the memory function $K_{SS}(\tau)$, Eq. (54), is proportional to the correlation function $C_x(\tau)$,

$$C_x(\tau) = \frac{\mathrm{Tr}\{(\sum_j B_j I_{xj}) e^{iP_2(\cos\theta_I)\cdot N_{II}\cdot\tau} (\sum_k B_k I_{xk})\}}{\mathrm{Tr}\{(\sum_j B_j I_{xj})^2\}} \tag{56}$$

where

$$B_j = \gamma_I \gamma_S \hbar (1 - 3\cos^2\theta_j) r_j^{-3} \cdot \frac{\mu_0}{4\pi}$$

is the structural factor of the dipolar coupling between the spins $I_j$ and $S$. $H_{II}$ is the secular part of the $I$–$I$ dipolar Hamiltonian. Suggesting at first that the $I$–$I$ dipolar coupling is static, $C_x(\tau)$ is proportional to the transverse relaxation function (the "FID") of the $I$ spin system where the time variable is scaled by the function

$$P_2(\cos\theta_I) = \tfrac{1}{2}(3\cos^2\theta_I - 1),$$

where $\tan\theta_I = \omega_{1I}/\Delta\omega_I$.

The form of $C_x(\tau)$ is found empirically. As mentioned before in studies of the polarization transfer in $CaF_2$, McArthur et al. [136] and Demco et al. [137] used a Gaussian function

$$C_x(\tau) = \exp\left(-\frac{\tau^2}{\tau_c^2}\right). \tag{57}$$

Finally,

$$\frac{1}{T_{IS}} = \frac{1}{2}\sin^2\theta_S \sin^2\theta_I \cdot M_2^{IS} J_x(\omega_{eS} - \omega_{eI}) \tag{58}$$

where $J_x(\omega)$ is calculated by inserting Eq. (57) into Eq. (55):

$$J_x(\omega) = \frac{\tau_c}{2}\sqrt{\pi} e^{-\omega^2\tau_c^2/4}.$$

The calculation satisfactorily reproduced the dependence of the experimental $T_{IS}^{-1}$ of $CaF_2$ (CP from the abundant $^{19}F$ to rare $^{43}Ca$ spins) on the irradiation strength, the mismatch of the coupling condition, and the irradiation frequency compared to the resonance frequency of the $S$ spin. To calculate the cross-polarization rates $T_{IS}^{-1}$, second and higher order approximations are used to determine $C_x(\tau)$. If only a second order approximation is rationalized, the correlation time introduced in Eq. (57) is expressed by the second moment $N_2$ of the correlation function $C_x(\tau)$ according to Eq. (56), i.e.

$$\tau_c^2 = \tfrac{1}{2} N_2. \tag{59}$$

The corresponding formulae are well-known [130]; $N_2$ contains the lattice sums

$S_1, S_2, S_3$

$$N_2 = \frac{1}{24}\frac{I(I+1)\gamma_I^4\hbar^2\mu_0^2}{(4\pi)^2}(P_2(\cos\theta_I))^2\cdot\frac{5S_3+8S_2}{S_1}, \tag{60}$$

with

$$S_1 = \sum_j b_j^2, \quad S_2 = \sum_{i\neq j} a_{ij}^2 b_i b_j, \quad S_3 = \sum_{i\neq j} a_{ij}^2(b_i^2 + b_j^2)$$

where the structural factors are given by

$$b_j = (1 - 3\cos^2\theta_j)r_j^{-3}(I\text{–}S), \quad a_{ij} = (1 - 3\cos^2\theta_{ij})r_{ij}^{-3}(I\text{–}I).$$

For the *polycrystalline sample* $N_2$ can be expressed by the Van Vleck second moment $M_2^{II}$ of the abundant spins,

$$N_2 = \tfrac{5}{36}M_2^{II}. \tag{61}$$

A simple expression for $T_{IS}$ results if the two r.f. fields are applied exactly at resonance for $I$ and $S$ spins ($\theta_I = \theta_S = \pi/2$) and if the Hartmann-Hahn condition is fulfilled, i.e. if the spectral density function in Eq. (58) reduces to $J_x(0) = \frac{1}{2}\tau_c\cdot\sqrt{\pi}$,

$$\frac{1}{T_{IS}} = \frac{3}{2}M_2^{IS}\cdot\left(\frac{2\pi}{5M_2^{II}}\right)^{1/2} \cong 1.681\cdot\frac{M_2^{IS}}{(M_2^{II})^{1/2}}. \tag{62}$$

$M_2^{II}$ and $M_2^{IS}$ are the Van Vleck second moments (in units of $s^{-2}$) for $I$–$I$ and $I$–$S$ dipolar couplings.

This rather simple formula for $T_{IS}^{-1}$ has been used to derive information on the local structure around the rare $S$ spins (i.e. interpretation of $M_2^{IS}$) in polycrystalline material from the analysis of cross-polarization rate in combination with lineshape data (i.e. determination of $M_2^{II}$) for the abundant spins. Such selective measurements were reported in the case $^{19}F \rightarrow {}^{13}C$ CP transfer for $CF_3COOAg$ [130], and for polymers [140] and in case of $^{19}F \rightarrow {}^{29}Si$ CP transfer for glasses [141, 142].

## 2.4 Cross-Polarization for Mismatch Hartmann-Hahn Conditions. Modified Spin Thermodynamic Theory

### 2.4.1 Introduction

In the preceding discussion, cross-polarization was described as a thermal equilibration between the $S$ spin and $I$ spin reservoirs under the influence of $I$–$S$ coupling. Hence, the evolution of the spin density operator is assumed to proceed in two steps: A short-time-scale process leads to an internal thermal equilibrium in each reservoir described by the corresponding spin temperatures, and a long-time-scale evolution leads to a uniform temperature of the two reservoirs. The only constraint was that the total energy is conserved. One prediction of this simple thermodynamic model is that the spin polarization increases proportional with the capacity of the $S$ spin reservoir and, hence with the amplitude $B_{1S}$ of the r.f. field applied with the Larmor frequency of the $S$ spins. This prediction is in contrast to experimental results which reveal that optimum cross-polarization is achieved under Hartmann-Hahn match. The discrepancy is usually explained by purely kinetic arguments, i.e. by a rapid decrease of the CP rate away from match so that in fact a quasi-equilibration is not attained.

Thus, it is necessary to include into the spin thermodynamics the effect of "broadening" of the Hartmann-Hahn match, i.e., a distribution of energy levels by dipolar interactions.

### 2.4.2 *Modified Spin Thermodynamic Theory*

The aim of the modified theory [133] is to introduce the intermediate time-scale phenomenon of quasi-equilibrium in which only partially a thermal equilibrium between the thermodynamic reservoirs occurs and which is a consequence of the quantization of the spin-systems. It has no parallel in macroscopic "classical" thermodynamics.

Quasi-equilibrium may be established if at least three reservoirs exist. The question arises [133] how to take into account the thermodynamic effect of quantization of the Hamiltonian $H_I$ and $H_S$ and the involvement of dipolar interactions. The theory presented by Levitt et al. [133] suggests that mismatched cross-polarization of a spin $S$ affects strongly the dipolar $I$–$I$ energies of a certain number of $I$ spins in the close vicinity of the $S$ spin but these energetic effect diffuse only slowly to those $I$ spins being more distant from the cross-polarization site. The spin thermodynamics in the case of three thermodynamic reservoirs assumes first that on a time scale $t < \tau_{\text{intern}}$ an internal thermal equilibrium is established individually within each reservoir. Over a time $\tau_{\text{que}}$ longer than $\tau_{\text{intern}}$ but shorter than $\tau_{\text{eq}}$, a quasi-equilibrium state may be reached in which only partial thermal equilibrium between the thermodynamic reservoirs occurs and which is a consequence of the quantization of the spin systems. This is a metastable state with nonuniform spin temperature and with less than maximum entropy. After a sufficiently long time $\tau_{\text{eq}}$ a complete thermal equilibrium may be attained.

An appropriate framework for solving such problems has been supplied by Jeener [143] in terms of orthogonal quasi-invariant operators whose eigenvalues may be assumed to be conserved during the process of thermal equilibration. As described in Sect. 2.1, we separate the Hamiltonian of the spin system into a set of thermodynamic reservoir terms denoted by $H_i^0$ and a smaller perturbation term $H^1$,

$$H = H^0 + H^1, \quad H^0 = \sum_i H_i^0 \tag{63}$$

with the commutators

$$[H_i^0, H_j^0] = 0 \quad \text{and} \quad [H^0, H^1] \neq 0. \tag{64}$$

We assume now that a set of orthogonal quasi-invariant operators $Q_q$ can be found, i.e.

$$(Q_q|Q_{q'})=0 \tag{65}$$

unless $q=q'$. Furthermore, it is requested that each operator $Q_q$ may be expressed as a linear combination of the terms $H_i^0$. If the operators $Q_q$ are normalized such that

$$(H^0|Q_q)=(Q_q|Q_q), \tag{66}$$

we find

$$H^0=\sum_q Q_q \tag{67}$$

and in comparison with the Eq. (63) the terms $Q_q$ may be interpreted as new thermodynamic reservoirs. Under different experimental conditions different choices for the quasi-invariant operators may be appropriate which may be also operative on different time scales.

For a CP process near Hartmann-Hahn match it is possible to assume [133] that exchange of order between $H_I^T$ and $H_S^T$ (cf. Eq. 10) takes place by spin-pair flips under the influence of the operators $I^+S^-$ and $I^-S^+$ contained in $H^1=H_{IS}^T$. Since the operator of the total spin momentum,

$$\sum_{k=1}^{N} I_{kz}+S_z=L_z$$

(abbreviated here as $L_z$), commutes with $I^+S^-$, $I^-S^+$ and the other Hamiltonians $H_I^T$, $H_S^T$, and $H_{II,sec}^T$ and since $L_z$ is clearly conserved by spin-pair flips, the first quasi-invariant operator $Q_1$ may be assigned to $L_z$:

$$Q_1=q_1\cdot L_z, \tag{68}$$

with the normalization factor

$$q_1=\frac{(H^0|L_z)}{(L_z|L_z)}=\frac{N\omega_{1I}+\omega_{1S}}{N+1}. \tag{69}$$

Here $N$ is the number of abundant spins $I$ which are in contact with a particular rare spin $S$ on an experimental time scale. If resolved, tight $I$–$S$ dipolar couplings are absent, we may assume that $Q_1$ is the only quasi-invariant of the system apart from the total energy [133]. Consequently, a quasi-invariant operator $Q_2$ can then be constructed simply by

$$Q_2=H^0-Q_1, \tag{70}$$

which together with Eqs. (67) and (68) leads to

$$Q_2=\frac{\Delta\omega}{N+1}\left\{NS_z-\sum_{k=1}^{N} I_{kz}\right\}+H_{II,sec}^T \tag{71}$$

where $\Delta\omega=\omega_{1S}-\omega_{1I}$.

In order to evaluate the achievable $S$ spin polarization under CP conditions it is necessary to construct the quasi-equilibrium spin density operator $\sigma^{\mathrm{T}}(\tau_{\mathrm{que}})$ and to calculate

$$M_{\mathrm{S}} = N_{\mathrm{S}}\gamma_{\mathrm{S}} \cdot \mathrm{Tr}\{S_z \sigma^{\mathrm{T}}(\tau_{\mathrm{que}})\}. \tag{72}$$

The quasi-equilibrium position can be deduced when the density operator of the *internal equilibrium* realized at a time scale $\tau_{\mathrm{int}}$ is projected separately on each quasi-invariant operator

$$\sigma(\tau_{\mathrm{que}}) \cong \sum_{\mathrm{q}} \frac{(\sigma(\tau_{\mathrm{int}})|Q_{\mathrm{q}})Q_{\mathrm{q}}}{(Q_{\mathrm{q}}|Q_{\mathrm{q}})} = Q_1 \frac{(\sigma(\tau_{\mathrm{int}})|Q_1)}{(Q_1|Q_1)} + Q_2 \frac{(\sigma(\tau_{\mathrm{int}})|Q_2)}{(Q_2|Q_2)}. \tag{73}$$

The spin density operator $\sigma(\mathrm{int})$ may be chosen to be the spin density operator $\sigma^{\mathrm{T}}(0)$ describing the spin system immediately after the initial $90^\circ_y$ pulse applied on the $I$ spin system being before in thermal equilibrium at the lattice temperature $T$. The spin density operator $\sigma_0(0)$ in the "tilted" frame is

$$\sigma^{\mathrm{T}}(0) = \frac{\hbar\gamma_{\mathrm{I}} B_0}{kT\,\mathrm{Tr}\{1\}} \sum_{\mathrm{k}} I_{\mathrm{kz}} = \alpha_{0\mathrm{I}} \sum_{\mathrm{k}} I_{\mathrm{kz}}. \tag{74}$$

Finally, inserting Eq. (74) into Eq. (73), and assuming $N \gg 1$, the quasi-equilibrium spin density operator is then calculated to be

$$\sigma^{\mathrm{T}}(\tau_{\mathrm{que}}) \cong \alpha_{0\mathrm{I}} \left\{ S_z \frac{\lambda^2}{\lambda^2 + \Delta\omega^2} + \sum_{\mathrm{k}} I_{\mathrm{kz}} + H^{\mathrm{T}}_{\mathrm{II,sec}} \frac{\Delta\omega}{\lambda^2 + \Delta\omega^2} \right\} \tag{75}$$

with the cross-polarization width $\lambda$,

$$\lambda^2 = \frac{(H^{\mathrm{T}}_{\mathrm{II,sec}}|H^{\mathrm{T}}_{\mathrm{II,sec}})}{(I_{\mathrm{kz}}|I_{\mathrm{kz}})} = \frac{1}{4} N M^{\mathrm{II}}_2 \tag{76}$$

and with the Hartmann-Hahn mismatch

$$\Delta\omega = \omega_{1\mathrm{I}} - \omega_{1\mathrm{S}}. \tag{77}$$

Using this spin density operator $\sigma^{\mathrm{T}}(\tau_{\mathrm{que}})$ we may derive the achievable $S$ spin magnetization $M_{\mathrm{S}}$ in the form

$$M_{\mathrm{S}} = \frac{N_{\mathrm{S}}\gamma_{\mathrm{I}}\gamma_{\mathrm{S}}\hbar^2 S(S+1) B_0}{3kT} \cdot \frac{\lambda^2}{\lambda^2 + (\Delta\omega)^2} \tag{78}$$

or, introducing the Curie formula for the $S$ spin polarization (magnetization) in thermal equilibrium $M_{0\mathrm{S}} = (C_{\mathrm{S}} B_0)/T$ as defined in Eq. (21)

$$M_{\mathrm{S}} \cong M_{0\mathrm{S}} \cdot \frac{\lambda^2}{\lambda^2 + (\Delta\omega)^2}. \tag{79}$$

In conclusion, the modified thermodynamic model of heteronuclear cross-polarization

- approximately accounts for the gain of magnetization proportional to $\gamma_{\mathrm{I}}/\gamma_{\mathrm{S}}$ for a single CP contact

– predicts that the achievable $S$ spin polarization is at a maximum on Hartmann-Hahn match $\Delta\omega = 0$, and falls off for both higher and lower r.f. field ($\sim \omega_{1S}$) as a Lorentzian shaped function of half-width at half-height $\lambda = \frac{1}{2}(NM_2^{II})^{1/2}$. $M_2^{II}$ is the Van Vleck second moment of the $I$–$I$ spin dipolar coupling.

The Lorentzian shape of the matching curve with a maximum at $\Delta\omega = 0$ is only qualitatively in accordance with the experiment. Two aspects have to be mentioned in this respect, viz. (i) that usually a Gaussian dependence of the CP transfer rate on the Hartmann-Hahn mismatch is found (see Sect. 2.3) and (ii) that quantitative agreement for the halfwidth $\lambda$ would require a suitable choice of the number $N$ of $I$ spins participating in the CP process with a single $S$ spin on an experimental time scale. The experimental situation for the CP half-width is usually such that $N$ is much less than the ratio of the abundances of the species $I$ and $S$ involved in the CP process (e.g. 150 for $^{1}H \rightarrow {}^{13}C$ CP). Hence, a proper estimate of the parameter $N$ in agreement with the experimental half-width $\lambda$ relies again on the fact that mismatch cross-polarization of an $S$ spin strongly effects the dipolar $I$–$I$ energy only in the close vicinity of the CP site and that energetic effects diffuse comparatively slowly to those $I$ spins being at larger distances from it.

Finally, modifications of the projection operators for resolved heteronuclear $I$–$S$ couplings [133] enable also the treatment of the experimental situation where the cross-polarized magnetization evolves in a damped oscillatory fashion, in particular close to Hartmann-Hahn match, $\Delta\omega \simeq 0$. These dipolar oscillations could not be explained in the framework of the simple thermodynamic theory presented in Sect. 2.2.

### 2.4.3 *Cross-Polarization Experiments Under Mismatched Hartmann-Hahn Conditions*

*Mismatch-optimized IS transfer (MOIST) [133].* As has been shown above, only a certain percentage of rare $S$ spins in the sample are polarized under mismatched Hartmann-Hahn condition. The amount of polarization transferred in the case of mismatched cross-polarization may be increased by a double irradiation scheme shown in Fig. 6. It resembles the normal Hartmann-Hahn cross-polarization technique but instead of coupling the spin-lock fields at constant phases, simultaneous $\pi$ phase shifts are introduced on both channels. The intervals between the synchronous phase reversals are long enough to allow the system to attain a quasi-equilibrium before the next phase shift. The technique is effective for static samples if a strong heteronuclear coupling exists which can contain the heteronuclear dipolar order. The creation of dipolar order limits the amount of polarization which may be transformed in the case of mismatched r.f. fields. This mismatch sensitivity can be overcome by the periodically $\pi$ phase-shifted spin-lock r.f. fields as shown in Fig. 7 [144]. The experimental time dependence of the $S$ spin polarization during a MOIST experiment is shown in Fig. 7. On each phase reversal the system is displayed from equilibrium and

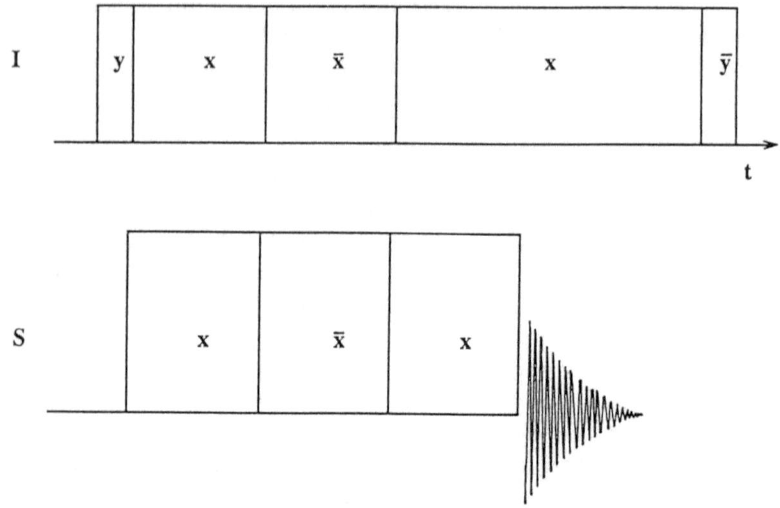


**Fig. 6.** Pulse sequence for the mismatch-optimized IS transfer (MOIST) experiment. In modifying the normal Hartmann-Hahn cross-polarization a synchronous phase reversal of the two spin-locking r.f. fields is introduced which is indicated by the steps $x$ in both channels

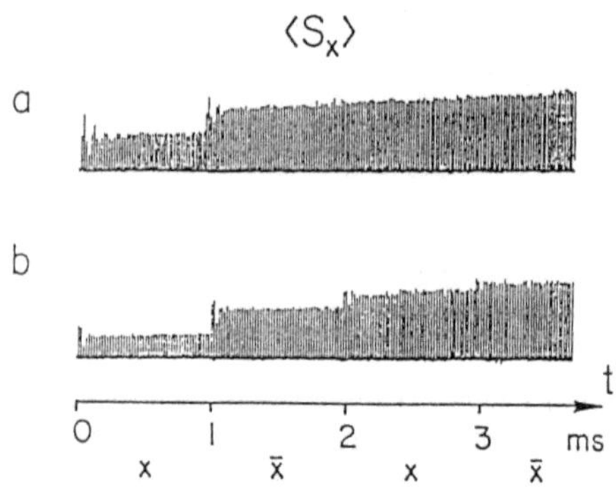


**Fig. 7a,b.** Evolution of the $S$ spin magnetization during a MOIST experiment (reproduced from Levitt, Suter, and Ernst [133] with permission of the American Institute of Physics). The phase of the two spin-locking r.f. fields are reversed every 1 ms as indicated by the phase $x$ and $-x$ below the time scale. The $S$ spin magnetization is monitored in steps of 20 μs. Different values of $\omega_{1S}/2\pi$ were used: **a** 38 kHz, **b** 31 kHz. Initial transient oscillations after a phase reversion are seen before the system reaches a new quasi-equilibrium with a large $S$ spin magnetization

after some period of transient oscillations (because of resolved $I$–$S$ coupling in ferrocene) a new quasi-equilibrium is attained with a larger $S$ spin polarization. The effect of the phase shift may be understood as an inversion of all Zeeman reservoir terms in $H_i^0$ and in the $Q_q$ (see Sect. 2.4.2). Since the dipolar terms remain invariant during the phase shifts the evolution of the system toward a new equilibrium position is initiated.

*Successive polarization under mismatched Hartmann-Hahn conditions [145].* A multiple contact cross-polarization pulse scheme shown in Fig. 8 is employed under mismatched Hartmann-Hahn condition. The essential point for the

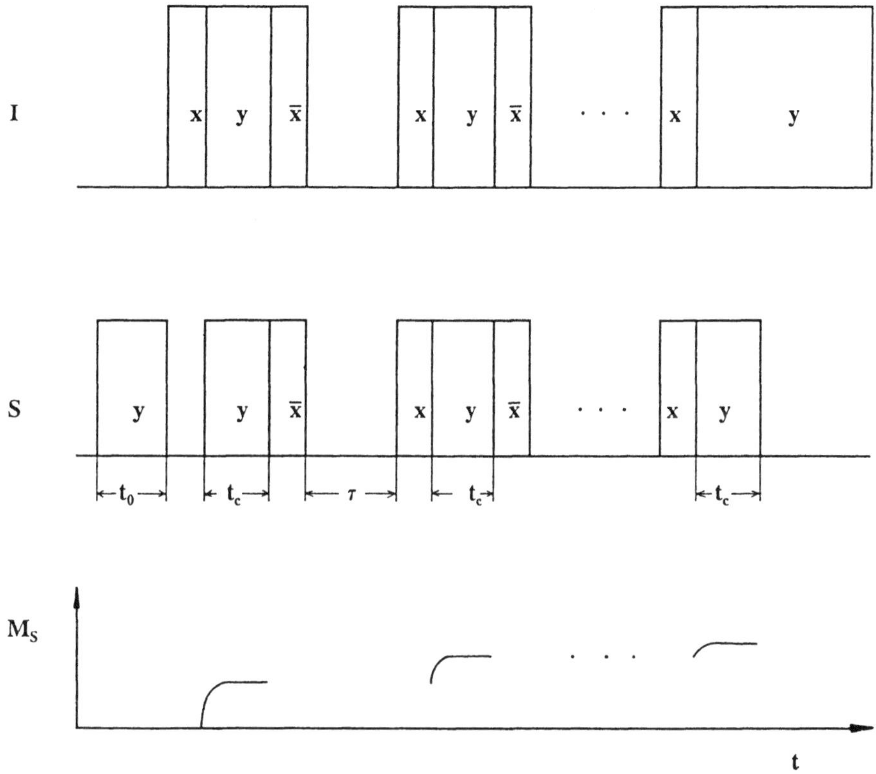


**Fig. 8.** Multiple contact cross-polarization scheme under mismatched Hartmann-Hahn conditions

application of this technique is the fact [133] that away from Hartmann-Hahn match the $S$ spin polarization $\langle S_z \rangle^T$ grows initially ($10 \cdots 100\,\mu s$) at a rate which is not much different from that under matched conditions, but reaches a certain level for the rest of the time with a lower value for $\langle S_z \rangle^T$ than under matched conditions. The system appears to reach a quasi-equilibrium which under mismatched conditions is less favourable with respect to creation of $S$ spin polarization than under Hartmann-Hahn match. An explanation of this unfavourable quasi-equilibrium is possible in the framework of the modified thermodynamic theory (Sect. 2.4.2). At the end of each contact the magnetizations $M_S$ and $M_1$ are turned parallel to the external field $B_0$ by the $90^\circ_{-x}$ pulses applied simultaneously on the $I$ and $S$ channels. Thus, $M_S$ is stored during $\tau$, if $\tau$ is chosen to be short compared with the $S$ spin-lattice relaxation time $T_{1S}$. Otherwise, the time interval $\tau$ should be long against $T_{1I}$ to enable the recovery of $M_1$. Moreover, if $\tau$ could be chosen longer than the inverse of the spin diffusion rate it is ensured that the $S$ spin have reached a single temperature through spin diffusion between polarized and unpolarized spins. In the next contact a certain number of $S$ spins is again polarized and the same procedure occurs as already described. Since spin diffusion will not change the total magnetization it is not essential for the maximum value of magnetization which has been achieved. The $S$ spin polarization (magnetization) increases successively with the increasing number of contacts until the increase resulting from one or more contacts is equal to the loss if magnetization caused by $T_{1S}$ effects and the imperfection of the $90^\circ_x$ and $90^\circ_{-x}$ pulses at the $S$ channel. The contact time is limited by $T_{1\rho I}$. Because of the limitations mentioned the full on-match $S$ spins polarization is in general not attained.

*Polarization transfer via a non-Zeeman spin reservoir.* In the mismatch-optimized IS transfer (MOIST) the mismatch sensitivity of the Hartmann-Hahn CP method was overcome by periodically applied $\pi$ phase-shifted spin-lock r.f. fields. This technique prevented the creation of heteronuclear dipolar oraer which limited the amount of $I$ spin polarization which may be transformed to $S$ spin polarization by energy matching between the Zeeman energy reservoirs in the rotating frame. Recently a highly efficient method of polarization transfer between nuclear spin ensembles was described were non-Zeeman reservoirs play a central role (Non-Zeeman Polarization Transfer-NZPT). A scheme has been successfully used by Vieth and Yannoni [146] which is based on thermal mixing utilizing the non-Zeeman reservoir of the spins for intermediate storage of spin order. Spin order in this case is defined with respect to local fields which mainly originate from dipolar or hyperfine interaction (e.g. for electron-nuclear polarization transfer). Such transfer of spin order from the primary Zeeman reservoir to the non-Zeeman system was achieved by application of a single short r.f. pulse into the wing of the NMR line of the 'source' spins $I$ (Fig. 9).

This pulse is followed by a pulse on the "target" spins $S$. The initial amplitude of the $S$ spin magnetization (e.g. in a $^{1}H-^{13}C$ polarization transfer [146]), observed after switching off the $S$ spin r.f. field, reflects the energy content at the end of the $I$ spin pulse. The transient $S$ spin magnetization may be detected as usually under conditions of $I-S$ spin decoupling. No transfer is obtained with a single $I$ spin pulse applied at resonance. Irradiation in the other wing of the $I$ spin resonance line results in a reversal of the sign of the $S$ spin magnetization. Best results in a $^{1}H-^{13}C$ NZPT experiment was obtained when the pulsed r.f. field for the $^{1}H$ nuclei was applied with a off-resonant field at the turning point of the absorption profile. The dependence of the NZPT on the amplitude of the $S$ spin r.f. field is relatively weak. The features of this experiment resembles the mechanism of electron-nuclear cross-polarization (see also Sect. 3). An efficient transfer was also observed with a sequence of two phase-shifted *resonant* pulses applied to the $I$ spin system similar as for a spin-lock pulse sequence. The amplitude $B_{1I}$ for an optimum polarization transfer was about 0.7 times the full width at half height of the $^{1}H$ resonance line.

A non-Zeeman polarization transfer has been also realized by a slight variation of a Hartmann-Hahn cross-polarization experiment. Zhang, Stejskal, Fornes, and Wu [2.18] performed a Hartmann-Hahn type of cross-polarization with dipolar order in the rotating frame instead of Zeeman order as realized

a)

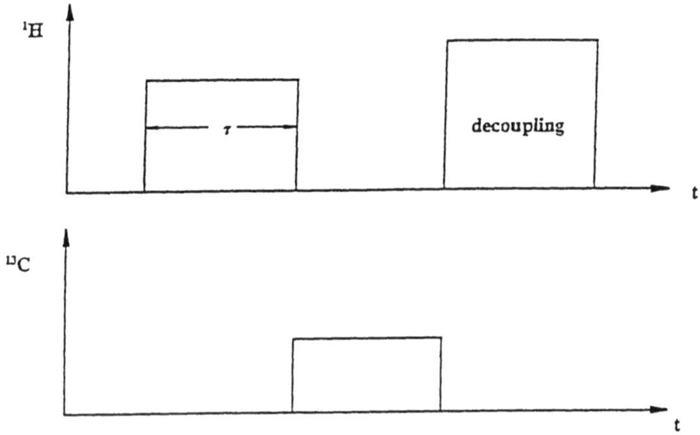


b)

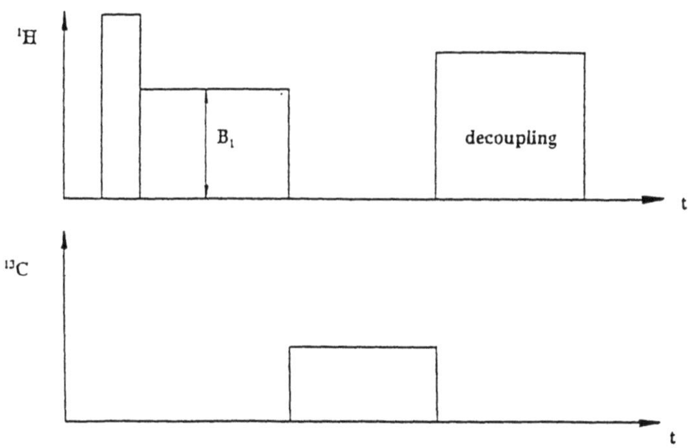


**Fig. 9a, b.** Transfer of spin order from the primary Zeeman reservoir to the non-Zeeman system by thermal mixing: **a** application of a *single* short r.f. pulse with duration $\tau$ into the wing of the NMR line of the source spins $I$, **b** sequence of two *phase-shifted resonant* pulses similar as used in the case of spin locking. The amount of $S$ spin magnetization transferred from the $I$ spin system varies with the amplitude $B_1$ of the r.f. field applied to the $I$ spin system.

in a usual spin-lock cross-polarization experiment. The pulse sequence is shown in Fig. 10. It startes with the Jeener-Broekaert $\pi/2$-$\pi/4$ pulse pair in the $I$ spin channel which convertes, as well-known, about 50% of the Zeeman order into dipolar order. During the time delay $t_2$ after the $\pi/4$ pulse, the residual Zeeman order is allowed to decay to zero. At the end of this time interval $t_2$ when the dipolar order is characterized by a common spin temperature, r.f. fields on both channels are applied such that cross-polarization may occur. A negligible cross-polarization takes place under Hartmann-Hahn match condi-

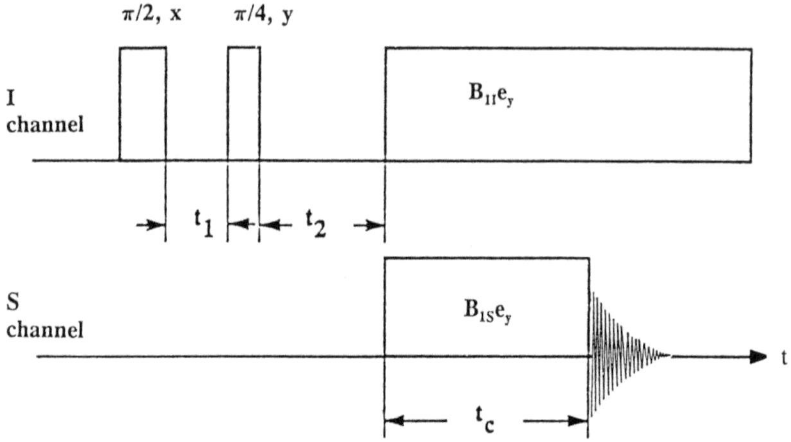


**Fig. 10.** Transfer of dipolar spin order from the *I* spin system to the system of *S* spins by means of cross-polarization. When during the CP experiment the dipolar spin order is not mixed into Zeeman order (ideal case), the *S* spin magnetization being transferred is zero for an exact Hartmann-Hahn match, i.e. $\omega_{1S} - \omega_{1I} = 0$. An optimum transfer is realized if $\omega_{1S} - \omega_{1I} = +\omega_D$ or $\omega_{1S} - \omega_{1I} = -\omega_D$ holds where positive or negative absorption signals appear for the *S* spin magnetization, respectively. The angular frequency $\omega_D$ of the local dipole interaction includes both the *I–I* and *I–S* contributions

tion $\omega_{1S} - \omega_{1I} = 0$, because in this case only energy levels in the rotating frame are involved having the same population. The maximum polarization transfer occurs if the condition $|\omega_{1S} - \omega_{1I}| = \omega_D$ is fulfilled where $\omega_D$ is proportional to the mean value of the *I–I* and *I–S* dipolar energy, viz. $\omega_D \propto \mathrm{Tr}\{(H_{II,sec} + H_{IS,sec})^2\}$, where $H_{II,sec}$ and $H_{IS,sec}$ are the secular parts of homonuclear and heteronuclear dipolar Hamiltonians, respectively. For $\omega_{1S} - \omega_{1I} > 0$ and $\omega_{1S} - \omega_{1I} < 0$ positive and negative spectra appear, respectively, if the r.f. fields are strong enough to overcome the dipolar fields and if no appreciable amount of dipolar order is directly transferred to the Zeeman order of the *I* spins during the CP period.

## 3 Cross-Polarization in Rotating Solids and in Systems with Thermal Motion

It was the combination of cross-polarization and magic-angle spinning (MAS) which led to the development of nowadays sophisticated high-resolution NMR techniques for solids. While in the majority of applications MAS is intended to average chemical shift anisotropy and to provide spectral information based on isotropic chemical shift, homonuclear and heteronuclear dipolar couplings are affected by MAS as well. The modulation of direct dipolar couplings originating from MAS, especially at high spinning rates ($\geqslant$ 10 kHz) and/or in spin systems with moderate or weak dipolar interactions, imposes profound effects upon the CP process itself.

In this section first a brief overview of the experimental phenomena encountered in conventional CP/MAS measurements is given (Sect. 3.1). Part of these observations can be understood by inspecting the transformation properties of the relevant dipolar coupling Hamiltonian, and on the basis of an extended statistical theory (Sect. 3.2). In order to overcome some of the difficulties in CP/MAS experiments as well as in their theoretical description, a whole new class of CP techniques has been proposed and these are reviewed in Sect. 3.3. Recently, experimental evidence led to the theoretical insight, that the assumption of a uniform spin temperature within one spin system, being in contact with another system during cross-polarization, might be not generally justified under high-speed MAS conditions and/or if Hartmann-Hahn mismatch occurs (Sect. 3.4). Finally, the effect of thermal mobility on cross-polarization is discussed (Sect. 3.5).

## 3.1 *Effects of Sample Spinning on Cross-Polarization*

The influence of MAS on the cross-polarization process can be illustrated best by considering the rare-spin signal intensity versus the Hartmann-Hahn mismatch $\omega_e = \omega_{eI} - \omega_{eS}$. The notation used here has already been defined in Sect. 2. In static samples, this Hartmann-Hahn matching curve is usually relatively broad and exhibits a Gaussian-like shape. With increasing spinning rates, the matching curve splits into a pattern of sidebands, separated by the spinning angular frequency $\omega_r$. The width and the intensity of each sideband also depends on $\omega_r$ and on the strength of the dipolar coupling between the abundant spins $I$. As long as the spinning angular frequency $\omega_r$ is much smaller than the $I$–$I$ dipolar coupling (expressed in angular frequency units) the matching curve is only weakly affected by MAS. This situation was first recognized by Stejskal, Schaefer, and Waugh [148] and evidence has been found by experiments on adamantane. Due to the relatively weak $^1H$–$^{13}C$ and $^1H$–$^1H$ dipolar couplings in this substance, the break up of the matching curve into a series of narrow sidebands occurs already at low MAS rates (2.3 kHz). For adamantane they found also a striking effect of sample spinning upon the CP rate $T_{IS}^{-1}$. The plot of $T_{IS}^{-1}$ versus the mismatch parameter $\omega_e$ revealed a similar sideband pattern as the matching curve. At $\omega_e = 0$, where maximum polarization transfer occurs for static samples, a lower value of $T_{IS}^{-1}$ was found in comparison to the value at $\omega_e = \pm\omega_r$ or $2\omega_r$ (first and second order sidebands). There is further experimental evidence for this effect. Dec, Wind, and Maciel [149] developed a high-speed MAS system, capable of spinning rates up to 23 kHz. Investigating the CP dynamics of Torlon under high-speed magic-angle spinning [150], they obtained matching curves for $^{13}C$ in protonated ($CH_n$) and non-protonated groups, which exhibit remarkable intensity deviations from each other at spinning rates above 10 kHz, reflecting the differences in CP efficiency for these molecular groups. Stejskal, Schaefer, and coworkers [151, 152] observed the spinning-sideband pattern of the CP rate versus matching in $^{15}N$–$^{13}C$–$^{1}H$ double cross-polarization NMR experiments under MAS conditions on amino acids and proteins at low spinning rates ($\leqslant 3.2$ kHz). Pruski et al. [153] studied the influence of MAS upon the $^{13}C$ CP/MAS NMR intensities of aromatic and aliphatic carbon in coal. The effect of spinning (at 5 kHz) on the signal intensity was found to be more pronounced for aromatic carbon, but not nearly as dramatic as for some model compounds (like, e.g., adamantane and hexamethyl-benzene).

The Hartmann-Hahn matching conditions in MAS and off-MAS experiments were investigated by Sardashti and Maciel [154] by $^{13}C$ and $^{113}Cd$ NMR for adamantane, non-deuterated and selectively deuterated cadmium formate and polyethylene. By deuteration it was possible to study the effect of $I$–$I$ dipolar coupling on the matching curve. While for adamantane and the deuterated compound (weak $^{1}H$–$^{1}H$ coupling) the matching curve splits into sidebands at spinning rates of 2.5 kHz, the matching curve is broad and featureless for polyethylene and non-deuterated cadmium formate (strong $^{1}H$–$^{1}H$ couplings). Spinning the sample around an axis inclined 90° relative to the external field instead the magic angle during the CP period showed also a drastic effect on the matching curve: (i) sidebands appeared separated by $2\omega_r$ and (ii) were significantly broadened compared to the matching cure under MAS (Fig. 11).

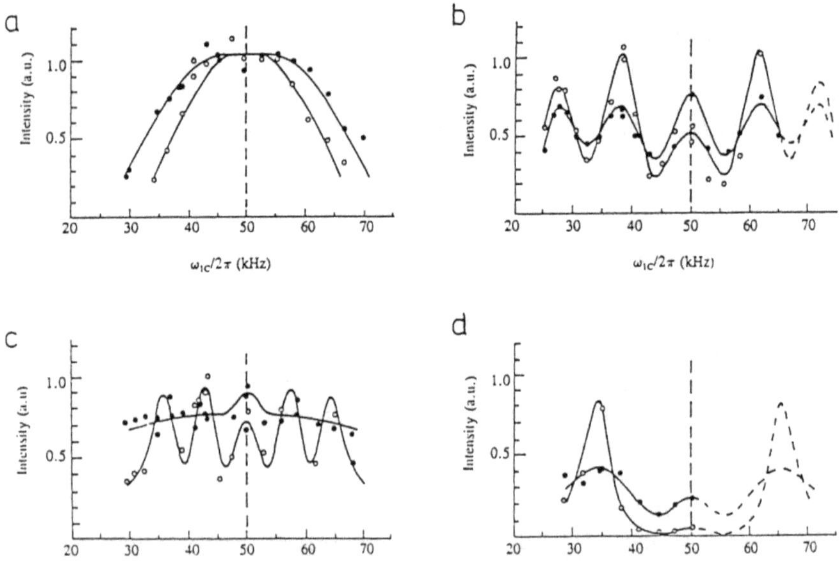


**Fig. 11 a–d.** Cross-polarization matching curves in rotating solids. The intensities of the $^{13}C$ NMR lines in protonated (*solid circles*) and nonprotonated aromatic $^{13}C$ nuclei (*open circles*) is plotted as a function of $\omega_{1S}$ (at constant $\omega_{1I}$) for different spinning frequencies $\omega_r/2\pi$: **a** 2.5 kHz; **b** 7.3 kHz; **c** 11.3 kHz; **d** 15.6 kHz (reproduced from Wind et al. [150], with permission of Academic Press, Inc. 1988)

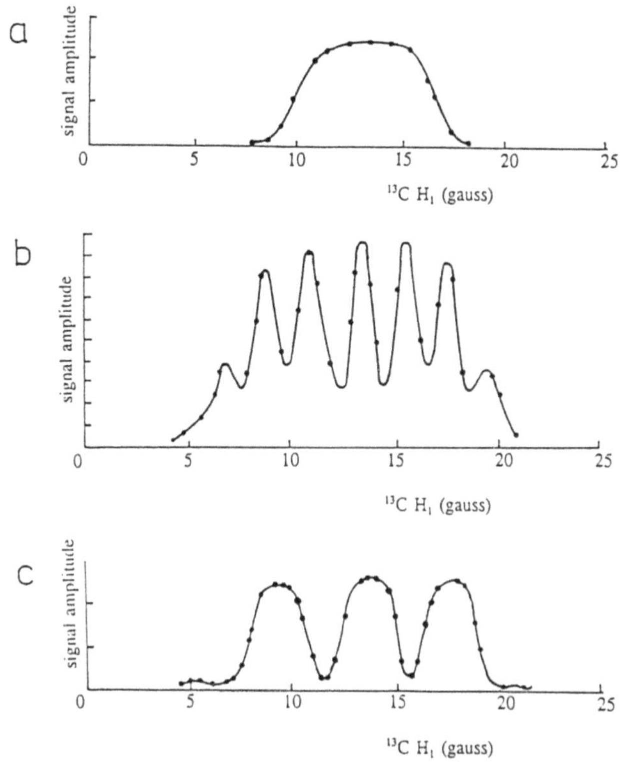


**Fig. 12a–c.** $^{13}C$ CP signal amplitude versus $^{13}C$ r.f. field (matching curve) for adamantane. **a** non-spinning sample, **b** MAS 2.5 kHz, **c** spinning at an axis inclined 90° to $B_0$ at 2.5 kHz (reproduced from Sardashti and Maciel [154], with permission from Academic Press, Inc. 1987)

Further experimental examples regarding matching-curve sideband patterns have been provided by Zeigler et al. [155], Barbara and Williams [156], Wu and Zilm [157], and Peersen et al. [158]. These results are reviewed in more detail in Sects. 3.3 and 3.4. The importance of the detailed appearance of the Hartmann-Hahn matching originates from two main reasons: (i) the matching curve provides information for a more complete theoretical understanding of the CP dynamics under sample spinning and mismatch conditions, and (ii) due to the occurence of narrow sidebands under MAS, matching conditions are much more critical under high-speed MAS than under static conditions, and are essential for spectrometer stability requirements and the reliability of quantitative conclusions drawn from intensities in CP/MAS spectra.

## *3.2 Theoretical Description*

The cross-polarization dynamics involves at least two conceptually different levels of description. The first includes the direct modulation of the dipolar interaction originating from sample spinning ("microscopic" level) and can be obtained unambiguously by examination of the transformation properties of the spin-Hamiltonians. The second level of description, much more intricate than the first, is concerned with cross-polarization as a thermodynamic equilibration process, involving a large number of spins, and includes explicit or implicit statistical assumptions about the spin systems.

### *3.2.1 Transformation of Dipolar Hamiltonians Under Sample Spinning*

The effect of sample spinning, a technique originally introduced by Andrew et al. [159], Lowe [160] and Kessemeier and Norberg [161], on the dipolar coupling between nuclear spins can be understood, considering the transformation properties of the corresponding Hamiltonians. The sample rotates about an fixed axis $R$, inclined by the angle $\theta$ with respect to $B_0$ and by the angle $\theta'_{ik}$ with respect to the $I$–$I$ and $I$–$S$ internuclear vectors $r_{jk}$ and $r_k$, respectively (Fig. 13).

The transformation of the secular parts of the $I$–$I$ and $I$–$S$ dipolar Hamiltonians,

$$H^0_{II} = \sum_j \sum_k d_{jk}[2I_{jz}I_{kz} - \tfrac{1}{2}(I_j^+ I_k^- + I_j^- I_k^+)]$$

$$H^0_{IS} = \sum_k b_k 2I_{kz}S_z, \tag{80}$$

$$d_{jk} = -\frac{\mu_0 \gamma_I^2 \hbar^2}{4\pi r_{jk}^3}\frac{1}{2}(3\cos^2\theta_{jk} - 1), \quad b_k = -\frac{\mu_0 \gamma_I \gamma_S \hbar^2}{4\pi r_k^3}\frac{1}{2}(3\cos^2\theta_k - 1)$$

may be easily achieved, since the coupling constants $d_{jk}$ and $b_k$ are both proportional to the spherical harmonic $(3\cos^2\theta_{jk} - 1)/2$. For the mechanical rotation with the angular frequency $\omega_r$ we can express $\cos\theta_{jk}$ by means of the trigonometric identity

$$\cos\theta_{jk} = \cos\theta'_{jk}\cos\theta + \sin\theta'_{jk}\sin\theta\cos(\omega_r t + \phi_{jk}) \tag{81}$$

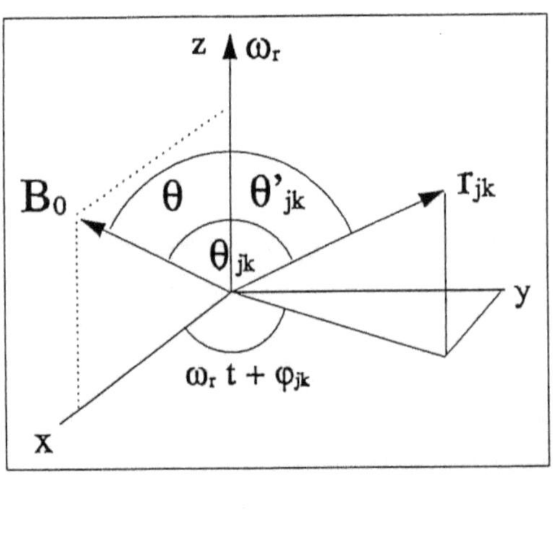


**Fig. 13.** MAS coordinate frame

and obtain

$$(3\cos^2\theta_{jk} - 1) = -\tfrac{1}{2}(3\cos^2\theta - 1)(3\cos^2\theta'_{jk} - 1) + \tfrac{3}{2}\sin 2\theta \sin^2\theta'_{jk}\cos(\omega_r t + \phi_{jk}) + \tfrac{3}{2}\sin^2\theta \sin^2\theta'_{jk}\cos(2\omega_r t + 2\phi_{jk}). \quad (82)$$

Equation (82) shows, that the sample rotation leads to a modulation of the dipolar coupling constants $d_{jk}$ and $b_k$ with frequencies $\omega_r$ and $2\omega_r$. The Hartmann–Hahn matching curve represents a function of these dipolar coupling parameters (*vide infra*), leading to the experimentally observed sideband pattern.

### 3.2.2 Statistical Description

As discussed in Sect. 2, the simple thermodynamic approach fails to describe cross-polarization in static samples under Hartmann-Hahn mismatch. The modified thermodynamic model proposed by Levitt et al. [133] allows the inclusion of CP mismatch conditions. The statistical theory of Demco, Tegenfeldt, and Waugh [137] (in the following referred to as DTW theory), although quite successful for the description of CP phenomena in static samples, is based on time-independent Hamiltonians in the r.f. rotating frame and in its original form does not account for sample spinning effects. A qualitative theoretical discussion of the main features of the CP/MAS matching curve on the basis of the weak-collision model used by Hartmann and Hahn [162] has already been outlined by Stejskal et al. [148].

A more detailed analysis of the CP/MAS matching curve has been proposed [163, 164], based on a modification of the DTW theory derived from Mori's and Zwanzig's projection technique [138, 139] (see also Sect. 2.3.). The description of the spin systems is based on a two-step spin-correlation function approach, introducing the functions $\gamma(t)$ for the $I$–$I$ dipolar coupling, and $C_x(t)$ for the $I$–$S$ polarization transfer. Both include the effect of sample rotation, assuming that spin fluctuations and sample rotation are *statistically* independent. As a consequence the two functions are decomposed into products, separating coherent MAS modulation expressed by the periodic function $K_{MAS}(t)$, and spin fluctuations, given by the transient functions $f(t)$ and $\gamma_F(t)$:

$$C_x(t) = M_{2IS} K_{MAS}(t) f(t), \quad \gamma(t) = M_{2II} K_{MAS}(t) \gamma_F(t). \quad (83)$$

$M_{2IS}$ and $M_{2II}$ denote the heteronuclear and homonuclear second moments of the dipolar interaction, respectively. The periodic function

$$K_{MAS}(t) = \frac{\langle\langle [1 - 3\cos^2\theta_{jk}(0)][1 - 3\cos^2\theta_{jk}(t)] \rangle\rangle}{\langle\langle [1 - 3\cos^2\theta_{jk}(0)]^2 \rangle\rangle} = \frac{2}{3}\cos(\omega_r t) + \frac{1}{3}\cos(2\omega_r t) \quad (84)$$

takes into account the *spatially averaged* time dependence of the local fields at the site of an $I$- or $S$-spin originating from sample spinning. The double brackets $\ll \cdots \gg$ denote the isotropic powder average. Under CP conditions $I$–$I$ and $I$–$S$ spin correlations are not independent of each other, expressed by the relation between the spin parts $f(t)$ and $\gamma(t)$:

$$\frac{d}{dt} f(t) = -f(t) \int_0^t \gamma(t')\,dt'. \tag{85}$$

Equation (85) can be derived from Mori's formalism [139] if the Anderson-Weiss approximation [165] is adopted, i.e. under the assumption that the $I$–$I$ spin correlations decay faster than the $I$–$S$ spin correlations. Postulating single exponential behaviour for the $I$–$I$ spin-correlation function $\gamma_F(t)$ in Eq. (83), $\gamma_F(t) = \exp(-t/\tau_F)$, for describing the spin dynamics in the absence of the sample rotation, allows an explicit calculation of the CP correlation function $C_x(t)$. The spin-correlation time $\tau_F$ depends on the strength of the dipolar coupling and is in the order of tens of microseconds for strongly coupled proton spin systems to more than 500 μs for weakly coupled systems (like in adamantane). If the short-correlation limit is valid, i.e. the characteristic time scale for the decay of $C_x(t)$ is much shorter than the time scale of the CP experiment itself then, analogous to DTW, the CP rate can be expressed as

$$T_{IS}^{-1} \sim \int_0^\infty C_x(t) \cos(\omega_e t)\,dt \tag{86}$$

with $\omega_e$ being the Hartmann-Hahn mismatch parameter as defined above.

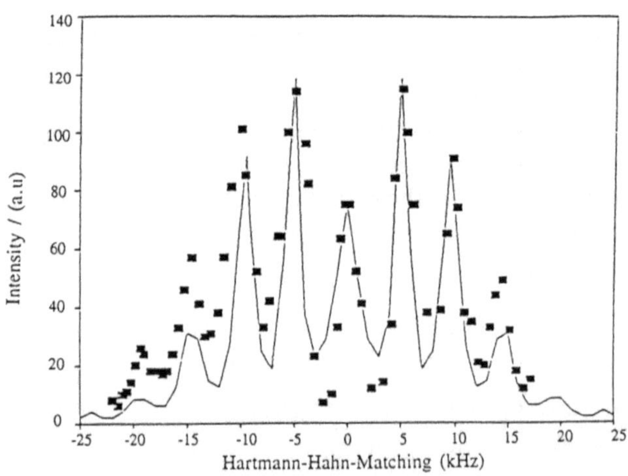


**Fig. 14.** Hartmann-Hahn matching for aromatic carbon in hexamethylbenzene, *squares*: experimental data [163], *full line*: calculated for a mixing time of 3 ms, spinning rate of 5 kHz, proton homonuclear second moment $M_{2II} = 1.7 \times 10^9\,s^{-2}$, $^{13}C$ proton heteronuclear second moment $M_{2IS} = 1.8 \times 10^7\,s^{-2}$ and a spin correlation time of $\tau_x = 600\mu s$

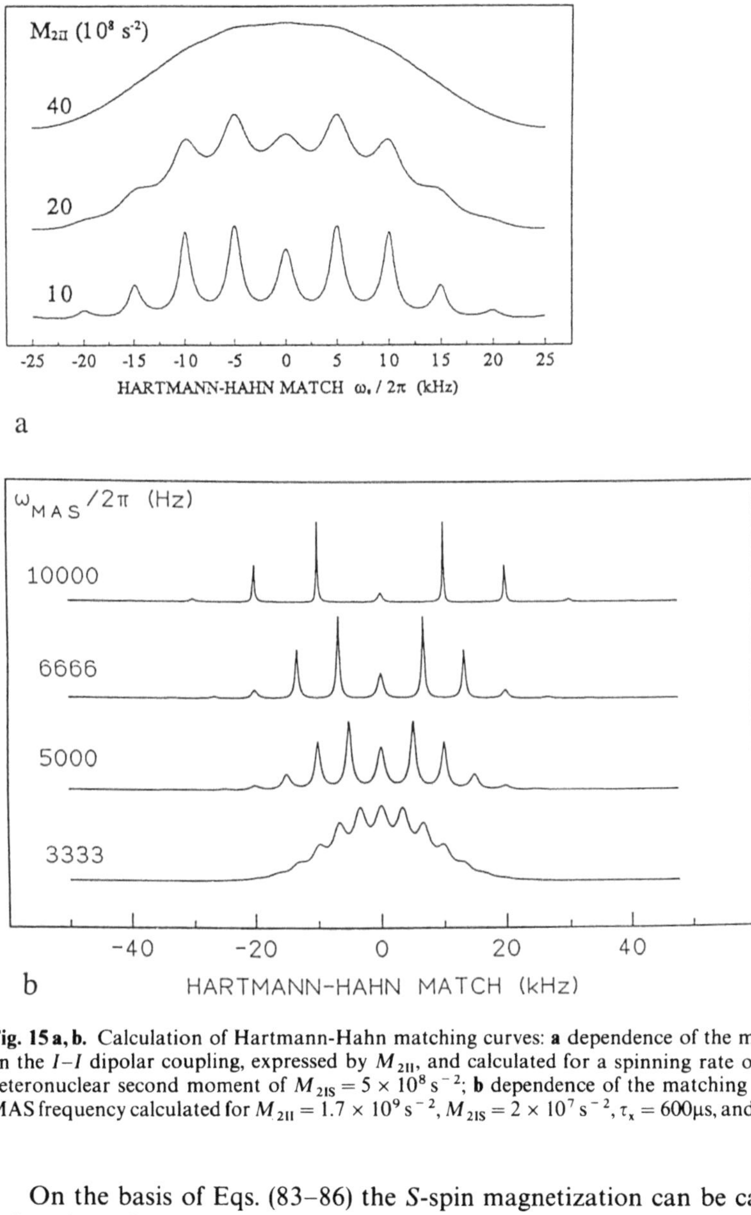


**Fig. 15 a, b.** Calculation of Hartmann-Hahn matching curves: **a** dependence of the matching curve on the *I–I* dipolar coupling, expressed by $M_{2II}$, and calculated for a spinning rate of 5 kHz and a heteronuclear second moment of $M_{2IS} = 5 \times 10^8\,s^{-2}$; **b** dependence of the matching curves on the MAS frequency calculated for $M_{2II} = 1.7 \times 10^9\,s^{-2}$, $M_{2IS} = 2 \times 10^7\,s^{-2}$, $\tau_x = 600\mu s$, and $t_{mix} = 1$ ms

On the basis of Eqs. (83–86) the *S*-spin magnetization can be calculated as a function of $\omega_e$, provided that the CP process is characterized by a single CP rate. A comparison of the theoretical matching curve with experimental $^{13}C$ CP/MAS NMR data for the aromatic carbon nuclei in hexamethylbenzene is shown in Fig. 14. Figure 15 illustrates the main features of the Hartmann-Hahn matching dependence on the strength of the *I–I* dipolar coupling. As soon as the spinning frequency $\omega_r$ becomes comparable to the homonuclear dipolar strength $M_{2II}^{1/2}$, the matching curve breaks up into sidebands, becoming narrower with further decreasing $I$–$I$ coupling. The same behaviour is found for a given constant value of $M_{2II}$ and increasing spinning rate.

Due to the inherent statistical assumptions, the theory discussed and illustrated above, is limited in its scope to spin systems with a more or less homogeneous distribution of $I$–$I$ and $I$–$S$ spin couplings, i.e. the spin systems can be characterized simply by their second moments of homo- and heteronuclear dipolar interactions and by correlation functions, which do not include locally different properties in the spin–spin couplings. These requirements are satisfied quite well by adamantane and hexamethylbenzene, where dipolar couplings are predominantly of intermolecular origin. In addition, the short-correlation limit in Eq. (86) and the Anderson-Weiss approximation represented by Eq. (85) restricts the validity to low or moderate MAS frequencies.

## *3.3 Techniques to Solve CP/MAS Matching Problems*

As discussed in Sect. 3.1, the combination of spin-lock cross-polarization and high-speed magic-angle spinning causes the CP matching curve to split into a pattern of sidebands, separated by the MAS frequency. As a result, the CP/MAS-spectral intensities become extremely sensitive to the spinning rate and stability conditions of the amplifier and the probe. Different strategies to overcome this problem have been proposed:

(i) *mechanical methods*, like off-magic-angle spinning [154] and stop-and-go spinning [155],
(ii) *r.f.-based techniques*, such as phase-modulated r.f. pulses [156, 157], variable amplitude cross-polarization [158], and amplitude-modulated cross-polarization [174].
(iii) alternative ways of polarization transfer via the *nuclear solid effect* and *dynamic nuclear polarization* [175, 176, 180].

### *3.3.1 Off-Magic-Angle Spinning* [154]

In order to avoid the experimentally undesirable situation of narrow CP-matching sidebands, Sardashti and Maciel [154] proposed off-magic-angle spinning (cf. also Sect. 3.1). During the cross-polarization period the sample is spun at a 90° angle instead the magic angle. The resulting CP-matching sidebands are separated by twice the spinning frequency and are significantly broader in comparison to CP/MAS matching sidebands. Following cross-polarization the $S$-spin magnetization is "stored" along the $z$ axis by a 90° pulse in order to gain the time interval necessary to flip the spinning axis to the magic angle. A final 90° pulse, applied after switching the spinner axis, and simultaneous $I$-spin decoupling allows detection of the $S$-spin signal.

### *3.3.2 Stop-and-Go Sample Spinning [155]*

An alternative method to solve the MAS interference upon the CP process has been developed by Zeigler, Wind, and Maciel [155]. During cross-polarization sample spinning is turned off or slowed down. After cross-polarization, similar to the off-MAS technique, the $S$-spin magnetization is rotated along the $z$ axis, and during this time the spinner is turned on. The final detection with simultaneous decoupling occurs under MAS. This technique allows the extension into a two-dimensional version, if immediately after the cross-polarization interval an evolution period $t_1$ is introduced, during which broadband $I$–$S$ dipolar decoupling is applied such that the $S$ spin magnetization evolves under the influence of complete chemical shift tensors. In the detection period $t_2$ the $S$-spin magnetization is influenced only by isotropic chemical shifts (MAS). Two-dimensional Fourier transformation with respect to $t_1$ and $t_2$ yields a 2D-NMR spectrum with chemical shift anisotropy displayed in the $\omega_1$ dimension, and isotropic chemical shift information available in the $\omega_2$ dimension.

### *3.3.3 R.F. Phase Modulation During CP [156, 157]*

Barbara and Williams [156] investigated the influence of rotor-synchronized, amplitude and phase modulated cross-polarization r.f. fields upon the CP matching curve. Applying phase reversal in both channels at twice the rotor period, the sideband structure of the CP matching curve for PMMA at a spinning rate of 10 kHz could be effectively suppressed and the matching curve became broadened similar to the static matching curve. However, they observed an undesirable reduction of the intensity by a factor of two. Without rotor synchronization, the phase modulated r.f. fields were ineffective in matching-sideband suppression. Wu and Zilm [157] independently obtained similar results

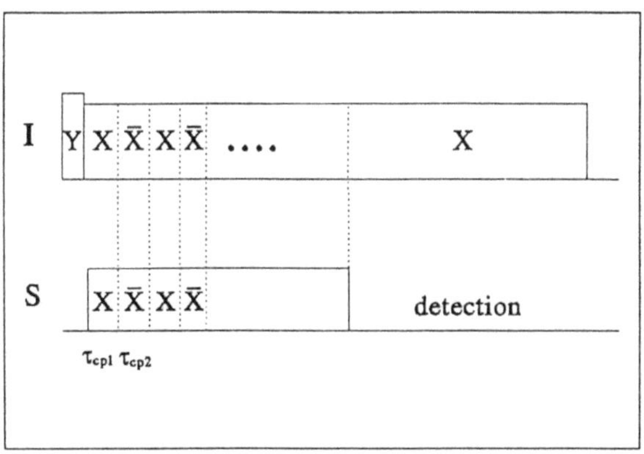


**Fig. 16.** Pulse sequence using simultaneous phase inversion during cross-polarization (SPICP)

using the simultaneous phase-inversion CP (SPICP) technique (Fig. 16). During CP the phase in both r.f. channels is phase shifted 180° simultaneously. This repeated phase inversion has to be synchronized with MAS for a Hartmann-Hahn mismatch of $\omega_{1I} - \omega_{1S} = \pm\omega_r$ (first sideband) in order to obtain an efficient CP gain, while for $\omega_{1I} \approx \omega_{1S}$ rotor synchronization is not important. The CP matching curve becomes broader if the frequency of phase inversions is increased. The scope of the SPICP technique is limited to solids with weak *I–I* dipolar couplings. The theoretical basis of the method is outlined in Sect. 3.4.2. Phase-inversion CP is closely related to the cross-polarization-depolarization scheme discussed in Sect. 3.4.1.

### 3.3.4 *Variable-Amplitude [158] and Amplitude-Modulated Cross-Polarization [174]*

The application of r.f. amplitude variations during the CP period (VACP) to manipulate the CP-matching sideband pattern has been successfully explored by Peersen, Wu, Kustanowich, and Smith [158]. The *I*-spin locking field amplitude is varied stepwise, symmetrically to the center of the matching curve, creating a series of slightly different Hartmann-Hahn matching conditions. The duration of each amplitude step has to be sufficiently long to allow a significant development of *S* magnetization during this step. By switching the r.f. amplitude 11 times during the spin-locking period in steps of 2.2 kHz with a duration of 455 μs per amplitude it was possible to suppress the matching sidebands and to maintain a broad matching curve for spinning rates up to 15 kHz for a sample consisting of a mixture of amino acids. As it is pointed out, within this technique, synchronicity of the r.f. amplitude changes (expressed in frequency units) with the rotor frequency has to be *avoided* in order not to create a CP-matching profile coinciding with the original, undesired matching curve. One drawback of this technique is the necessity of relatively long $T_{1\rho}$ relaxation times, because the CP period may become quite long due to the relatively high number of amplitude steps required.

Hediger, Meier, and Ernst [174] proposed *rotor-synchronized* amplitude-modulation cross-polarization (AMCP) to achieve higher CP rates at exact Hartmann-Hahn matching and/or a broadening of the CP-matching curve. There are two different types of pulse schemes: amplitude modulation in the *S*-spin channel (S-AMCP), and amplitude modulation in both channels (D-AMCP) (Fig. 17). The SPICP scheme [157] (Fig. 16) mentioned above can be regarded as a special D-AMCP sequence. While D-AMCP, similar as SPICP, is restricted in application to spin systems with relatively weak *I–I* dipolar coupling, S-AMCP is more generally applicable. However, S-AMCP does not solve the mismatch problem for the exact Hartmann-Hahn condition at high spinning rates.

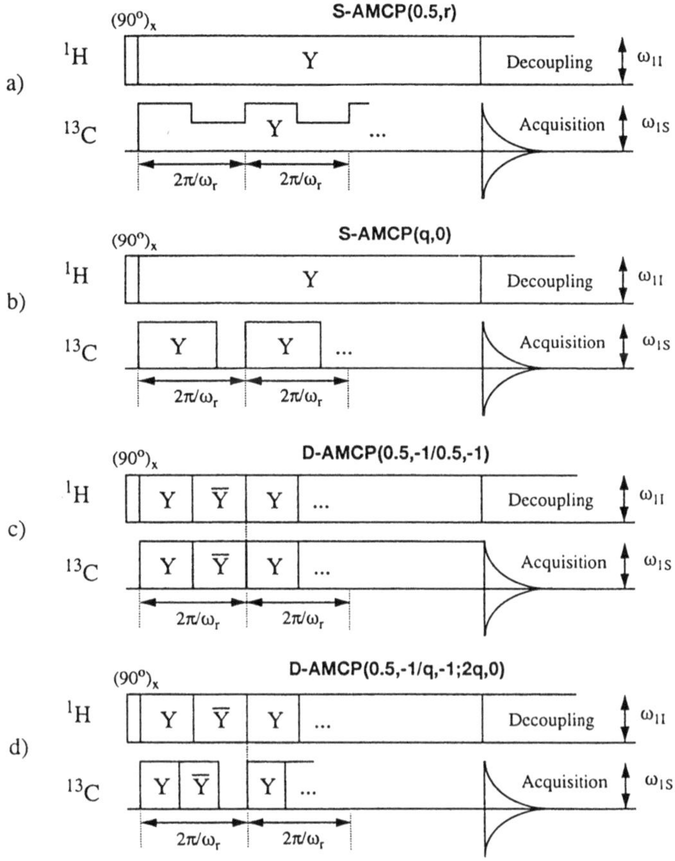


**Fig. 17.** Various amplitude-modulated CP schemes (AMCP) proposed by Hediger et al. [174] (reproduced from Hediger, Meier and Ernst [174] with permission from Elsevier, 1993). The denotions are as follows: **S-AMCP (q, r)** are CP sequences with MAS synchronized square modulation of the r.f. field at the $S$ spins while using a continuous r.f. field at the $I$ spins, where $q$ indicates the fraction of the MAS period after which the $S$ spin r.f. power level is changed by the ratio r. **D-AMCP ($\mathbf{q_I}, \mathbf{r_I}/\mathbf{q_S}, \mathbf{q_S}$)** are double modulated CP sequences with MAS synchronized square modulated $I$ spin and $S$ spin r.f. fields. Note that the CP time interval $2\pi/\omega_r$ is corrected in comparison with the original paper

### *3.3.5 Polarization Transfer via the Nuclear Solid Effect (NSE) [175,176] and Dynamic Nuclear Polarization (DNP) [180]*

The nuclear solid effect is analogous to the solid effect associated with dynamic nuclear polarization (DNP) [177]. In DNP, the nuclei are polarized by saturation of a so called "*forbidden transition*" involving simultaneous flips or flip-flops of an electron and nuclear spin. This transition occurs at the sum or difference of the electron and nuclear Larmor frequencies. In NSE, the abundant spins $I$ and rare spins $S$ play an analogous role as the electrons and nuclei in

DNP, respectively. The forbidden transition occurs at the sum $\omega_{0I} + \omega_{0S}$ or the difference $\omega_{0I} - \omega_{0S}$ of the Larmor frequencies of the participating nuclei $I$ and $S$. A combination of this well-established technique was proposed already in 1958 by Abragam and Proctor [177, 178]. With regard to modern spectroscopic methods it has been re-introduced by Wind and Yannoni [175, 176]. The NSE technique possesses several advantages over the conventional CP method of polarization transfer:

(1) no polarization loss due to insufficient locking of the proton magnetization,
(2) the $S$ spin polarization is independent of the proton relaxation rates (provided that thermal equilibrium of abundant spins is established before NSE transfer),
(3) easy optimization of the NSE match conditions since the transfer occurs via a matching of the frequency difference $\omega_{0I} - \omega_{0S}$,
(4) special advantage for rare spins with small $\gamma$ values for the same reason as mentioned under item (3),
(5) the spatial part of the $I$–$S$ coupling term responsible for NSE is *not influenced by magic-angle spinning* [179].

However, it should be mentioned that the only (but major!) disadvantage of the NSE method is that the transition probabilities rates $W_{\mp}$ (for the NSE transition at angular frequencies $\omega_{0I} \pm \omega_{0S}$) are much smaller than $T_{IS}^{-1}$, requiring strong r.f. irradiation to polarize the $S$ spins completely.

Analogous to the NSE method, the dynamic nuclear polarization (DNP) technique has been applied to enhance $^{13}C$ signals by one or two orders of magnitude in solids containing unpaired electrons [180], like coals, conductors or semiconductors, or other solids doped with radicals. Parent coal contains, for example, many free radicals (typically $10^{19}$ per gram). An experimental setup was designed to operate at a $B_0$ field of 1.4 T, corresponding to a proton frequency of 60 MHz, a $^{13}C$ frequency of 15 MHz and an electron frequency of 40 GHz. The main reason for using this relatively low field strength is related with the microwave equipment (because of the costs and the technical problems if operating at still higher frequencies). The DNP experiments have been combined with MAS to obtain *resolution enhancement in* $^{13}C$ NMR spectra. The *appreciable enhancement of signal intensities* due to DNP made it possible to obtain $^{13}C$ spectra of coal in a fast way, both directly via DNP-$^{13}C$ FID/MAS and indirectly via DNP-$^1H$ combined with $^1H$-$^{13}C$ CP/MAS experiments. DNP circumvents the limitations which MAS imposes on the CP process and may save spectrometer time in comparison to direct $^{13}C$ FID/MAS measurements, and not revealing the quantification problems for $^{13}C$ NMR intensities described above. Recently, $^1H$ DNP with CP/MAS $^{13}C$ NMR detection was applied to study polymer-blend interfaces [181].

## *3.4 Heterogeneity of Cross-Polarization*

### *3.4.1 Cross-Depolarization and Polarization Inversion*

Cross-polarization under mismatched Hartmann-Hahn conditions has been investigated not only focusing on sample spinning as discussed in Sect. 3.1. As mentioned in Sect. 2.4., Levitt et al. [133] introduced the quasi-equilibrium model into the CP theory. The evolution of the spin-density operator proceeds in three steps: A short time scale process leads to an internal thermal equilibrium in each reservoir, an intermediate time-scale process results in a quasi-equilibrium state between the *S* spin and the I spins in the close vicinity of it, and finally a long-time scale evolution establishes a uniform spin temperature of the total spin system.

Hahn et al. [166, 167] pointed out, in their detailed analyses, that the exponential law for the CP time dependence of the magnetization is not valid for the initial time evolution, if the relaxation is diffusion-limited. As already mentioned briefly in Sect. 2.3, Müller, Kumar, Baumann, and Ernst [3.21] observed strong transient oscillations for the $^{1}H$-$^{13}C$ magnetization transfer in ferrocene.

Nonexponential time-dependences of the magnetization transfer in cross-*depolarization* experiments have been reported recently by Wu et al. [169]. Cross-depolarization can be achieved (Fig. 18a) by inserting a time interval $\tau_i$ into the CP period, during which the *I*-spin locking field is switched off, causing the dephasing of transverse *I*-spin magnetization. The subsequent period $\tau_{c2}$ with both r.f. fields is then characterized by a magnetization transfer from the *S*-spin to the *I*-spin system, leading to a decrease of *S*-spin magnetization. After this depolarization time the *S*-spin signal is detected under *I*-spin decoupling.

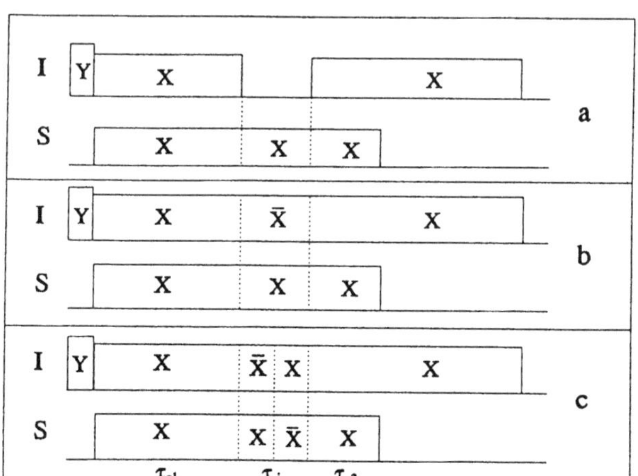


**Fig. 18a–c.** Pulse sequences suitable for cross-depolarization (CDP): **a** interrupted spin locking [169], **b** polarization inversion (CPPI) [172], and **c** simultaneous phase inversion (SPI) [172]

Monitoring the $^{13}C$ magnetization as a function of the depolarization time $\tau_{c2}$, a two-stage behavior was found for glycine and terephtalic acid: during the first 40 μs a fast decrease of $^{13}C$ magnetization was observed for $^{13}CH_2$ and $^{13}CH$ groups, followed by a slower decrease at longer times. The $^{13}C$ magnetization in carboxylic groups did not exhibit this feature, its decay appeared mono-exponentially.

The two-stage behaviour has been explained by taking into account, that $CH_2$ and CH groups form quasi-isolated $SI_n$ subsystems. The energy exchange within each subsystem is fast, leading to the initial fast depolarization. At longer times, also spin diffusion between different subsystems can take place, giving rise to the slow decay period observed.

Depolarization of $S$-spin magnetization can also be achieved by *selective polarization-inversion* (Fig. 18b, c) [170, 172] and utilized for spectral editing in $^{13}C$ CP/MAS NMR spectra. This technique has been independently proposed by Zumbulyadis [171] following a proposal from Melchior (see Ref. 12 in Ref. [171]). The cross-polarization period is divided into two time intervals. During the first period $t_c$, standard cross-polarization occurs. In the subsequent time interval, the r.f. phase of the $I$-spin locking field is 180° phase-shifted, while the phase in the $S$-spin channel remains unchanged. The phase switching in the $I$-spin channel leads to a negative $I$-spin temperature and a new equilibration process is initiated, leading to depolarization of the $S$-spin system. The $S$-spin depolarization develops with different depolarization rates according to the local proton environments for various $^{13}C$ nuclei ($CH_3$, CH, COOH), therefore providing a tool for editing $^{13}C$ CP/MAS NMR spectra similar to DEPT and INEPT for liquid-state NMR.

The cross-polarization/polarization-inversion (CPPI) technique for spectral editing purposes has been further explored and extended by simultaneous phase-inversion experiments (SPI) [172] in combination with magic-angle spinning at low as well as at high spinning rates. CH and $CH_2$ groups behave differently in the polarization-depolarization process. For the initial depolarization period (fast stage) these molecular groups can be treated as quasi-isolated $SI_n$ groups ($n = 1, 2$). A distinction between CH and $CH_2$ groups can be achieved, because the dipolar coupling Hamiltonians of these groups possess different properties under MAS (*vide infra*).

### *3.4.2 Simultaneous Phase Inversion CP (SPICP) and Rotor-Synchronized Amplitude-Modulation CP (AMCP) Schemes*

Phase-inversion CP methods, similar to the techniques considered above, have been explored as a means to circumvent Hartmann-Hahn mismatching originating from the split up of the matching curve into narrow sidebands at high MAS speed [156, 157]. However, because these approaches aim at an improved matching and polarization transfer under MAS and not at depolarization of the $S$ magnetization as described in Sect. 3.4.1, phase inversion has to take place in both channels ($I$ and $S$) simultaneously with equal r.f. phases in both channels (Fig. 16). A more general scheme, proposed by Hediger, Meier, and Ernst [174] designs pulse sequences on the basis of the theoretical description for general amplitude-modulated r.f. fields (Fig. 17). The main idea of using r.f. amplitude modulation, which includes simultaneous phase inversion (SPICP), to obtain broader matching curves or a higher CP rate at exact Hartmann-Hahn match under high-speed MAS, consists in making additional polarization transfer pathways accessible, which have normally a low transition rate in conventional continuous-wave cross-polarization (CWCP). The theoretical formalism proposed by Wu and Zilm [157, 172], and Hediger, Meier and Ernst [174] is bound to quasi-isolated $I$–$S$ spin pairs, i.e., the $I$–$I$ dipolar coupling is assumed to be weak and neither the $I$-spin system nor the $S$-spin system behave as thermodynamic reservoirs with a homogeneous spin temperature on the time scale of the CP experiment. If the amplitude modulation of r.f. fields $\omega_{1k}(t)$ $(k = I, S)$ is synchronized with the MAS angular frequency $\omega_r$, it is possible to express the r.f. fields, as well as the $I$–$S$ dipolar coupling strengths $b$, which are directly modulated by MAS, as Fourier series

$$\omega_{1k}(t) = \sum_{k} \omega_{1k}^{(n)} e^{in\omega_r t} \tag{87}$$

$$b(t) = \sum_{n} b_n e^{in\omega_r t}. \tag{88}$$

The corresponding Hamiltonian in the r.f. rotating frame, containing the r.f. fields and the dipolar coupling between $I$ and $S$ spins then becomes time-periodic with the MAS angular frequency $\omega_r$, and the solution of the Liouville-von Neumann equation for the spin-density matrix to obtain the cross-polarization dynamics for the $I$–$S$ spin system corresponds to solving linear first-order differential equations with time-periodic coefficients. One elegant possibility to solve the problem mathematically, is the application of Floquet's theory [182, 183], as proposed by Hediger et al. [174]. In the present case of the cross-polarization process, the Floquet theory describes the spin dynamics as governed by an average Floquet-Hamiltonian $H_F$ and Floquet states $|\alpha\beta, n\rangle$, where $|\alpha\beta\rangle$ denotes the product-spin Zeeman states (in the r.f. rotating frame) for the $I$–$S$ spin system and $|n\rangle$ being formally a "rotor state", given by $n$ as the Fourier indices introduced in Eqs. (87) and (88). The off-diagonal matrix elements of the Floquet-Hamiltonian $H_F$ (formed by the Floquet states) indicate, which cross-polarization transitions as functions of the mismatch $\omega_{1S}^{(0)} - \omega_{1I}^{(0)}$ and the "rotor state" $|n\rangle$ occur with an appreciable rate. It is shown, that without r.f.-amplitude modulation (CWCP, all $\omega_{1k}$ Fourier coefficients are zero except $\omega_{1k}^{(0)}$) cross-polarization mainly proceeds via $I$–$S$-flip-flop transitions, if the following conditions for Hartmann-Hahn match and selection rules are satisfied

$$\begin{aligned} &|\omega_{1S} - \omega_{1I} + n\omega_r| \leqslant \tfrac{1}{2} b_n \\ &n = \pm 1, \pm 2; \quad \Delta m_I = \pm 1, \quad \Delta m_S = \mp 1; \quad \Delta m = \Delta m_I + \Delta m_S = 0 \end{aligned} \tag{89}$$

expressing the fact, that under CWCP and high-speed MAS only the matching side bands of first and second order are important. Rotor-synchronized r.f.-amplitude modulation has the effect of introducing new off-diagonal matrix elements into $H_F$, which leads to the removal of the restriction $n = \pm 1, \pm 2$. Therefore, also cross-polarization at exact Hartmann-Hahn match ($n = 0$) becomes possible, in addition to the polarization transfer at the various side bands.

Two classes of amplitude-modulated CP (AMCP) experiments are possible. In *S*-AMCP pulse schemes only the *S*-spin r.f.-channel is square-wave modulated,

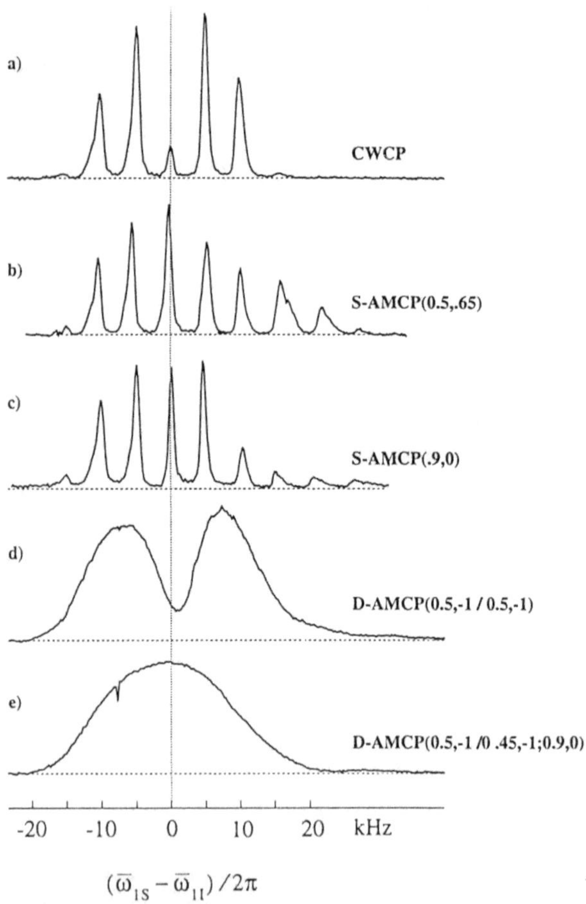


**Fig. 19a–e.** Cross-polarization matching curves for the $^{13}CH_2$ resonance line in adamantane obtained with the pulse schemes shown in Fig. 17 (reproduced from Hediger, Meier and Ernst [174] with permission from Elsevier, 1993). The standard Hartmann-Hahn cross-polarization is denoted by CWCP. The other abbreviations are introduced as in Fig. 17. For all experiments, the parameters chosen were: MAS frequency 5 kHz, mixing time 1.2 ms, $\omega_{1I}/2\pi = 22.7$ kHz. For each pulse sequence 256 experiments with variable $\omega_{1S}$ were recorded. The CP matching curves are plotted on the same amplitude scale and can be directly compared

while in the $I$-spin r.f.-channel the usual $I$-spin spin-locking conditions are maintained (Fig. 17a and b), allowing extended CP periods. However, the matching curve will contain narrow sidebands as in CWCP (Fig. 19a), but also a center band with a large amplitude (see Fig. 19b and c). Modulating the amplitude in both channels (Fig. 17c and d), called double amplitude-modulated CP (D-AMCP), results in a broadening of the matching curve (Fig. 19d and e). Beside the example for adamantane shown in Fig. 19 the efficiency of these techniques have been demonstrated for $^{13}C$ CP/MAS NMR spectra in polycarbonate [174], and hexamethylbenzene, ferrocene and glycine [157].

## *3.5 Influence of Thermal Mobility on Cross-Polarization*

### *3.5.1 Introduction*

In many applications of the CP NMR technique the influence of thermal mobility on the $I$–$S$ and $I$–$I$ dipolar coupling has to be considered. With increasing thermal mobility the secular terms of these coupling Hamiltonians become explicitly time-dependent and it is expected that the efficiency of cross-polarization is limited by thermal motion. Moreover, thermal motion may give rise to an inefficient heteronuclear decoupling and may lead to a limitation of resolution in CP NMR spectra measured under heteronuclear decoupling. In particular this situation may occur for intermediate values of the thermal correlation times $\tau_c \simeq \omega_{1I}^{-1}$ when efficient averaging due to the precession of magnetic moments in the r.f. field $B_{1I}$ is in competition with the broadening due to the thermal modulation of heteronuclear spin coupling (i.e. relaxational effects).

Two different approaches have been proposed to treat the influence of thermal mobility on the efficiency of the cross-polarization transfer:

(i) A "*short correlation time theory*" [184, 185] for nuclear magnetic relaxation in the presence of r.f. fields applied near the Larmor frequencies $\omega_{0I}$ and $\omega_{0S}$ of the spin system $I$ and $S$, respectively. Isotropic thermal motion is assumed. Starting from the limit of short correlation times, a system of two coupled linear differential equations for the $I$ and $S$ spin magnetizations along the instantaneous directions of the r.f. fields $B_{1I}$ and $B_{1S}$ is found in which the rate constants are given by a superposition of spectral density functions which describe the influence of thermal mobility. These equations have a structure which shows some analogy to the coupled equations derived in the framework of the simple thermodynamic theory in Sect. 2.2, Eq. (36). The maximum polarization transfer from the abundant $I$ spins to the rare $S$ spins is evaluated as a function of the thermal correlation time and these predictions are *extrapolated to the limit of long correlation times.*

(ii) In order to prove the validity of the "short correlation time theory" to be extrapolated, in a second approach the cross-polarization transfer is evaluated for the inverse situation [186], i.e. beginning from CP dynamics in solids and extrapolating the results to the case of intermediate correlation times. This treatment is based on an operator formalism proposed by Vega and Pines for the treatment of the interaction of a spin system (with quantum number $S=1$) with a weak r.f. field [187, 188].

The predictions of both approaches for the region or correlation times which are typical for the transition between solids and liquids coincide. In the following section only the theory for the short correlation limit is discussed in some more detail.

### 3.5.2 Short Correlation Time Limit

We suppose again a system an abundant spins $I$ (number $N_I$, magnetogyric ratio $\gamma_I$, $I=\frac{1}{2}$) and rare spins $S(N_S, \gamma_S, S=\frac{1}{2})$ in a d.c. field $B_0\bar{e}_z$ and under the influence of two perpendicular resonant r.f. fields $B_{1j}(t)=2B_{1j}\cos(\omega_{0j}t)$ $(j=I,S)$.

The rate equations for the components of magnetization $M_j$ aligned along the instantaneous directions of $B_{1j}(t)$ have been derived by Ernst, Fenzke, and Pfeifer [184] on the basis of Redfield's theory of relaxation:

$$\frac{dM_I}{dt} = -(R_{IO}+R_{II})\cdot M_I + R_{IS}\cdot M_S$$
$$\frac{dM_S}{dt} = +R_{SI}\cdot M_I - (R_{SO}+R_{SS})\cdot M_S. \tag{90}$$

These equations contain the following rate constants which depend both on the Larmor frequencies and the nutation frequencies in the r.f. fields, as well as on the correlation time of thermal motion. The rate constants $R_{IO}$ and $R_{SO}$ are given by

$$R_{IO} = \frac{N_S}{N_I}\cdot M_2^{IS}\cdot\{\tfrac{1}{2}J(\omega_{1I}-\omega_{1S})+\tfrac{1}{2}J(\omega_{1I}+\omega_{1S})+\tfrac{3}{2}J(\omega_{0S})$$
$$+\tfrac{3}{4}J(\omega_{0I})+\tfrac{1}{4}J(\omega_{0I}-\omega_{0S})+\tfrac{3}{2}J(\omega_{0I}+\omega_{0S})\}, \tag{91}$$
$$R_{SO} = M_2^{IS}\cdot\{\tfrac{1}{2}J(\omega_{1I}-\omega_{1S})+\tfrac{1}{2}J(\omega_{1I}+\omega_{1S})+\tfrac{3}{2}J(\omega_{0I})$$
$$+\tfrac{3}{4}J(\omega_{0S})+\tfrac{1}{4}J(\omega_{0I}-\omega_{0S})+\tfrac{3}{2}J(\omega_{0I}+\omega_{0S})\}.$$

Similar formulae hold for $R_{II}$, $R_{IS}$, and $R_{SI}$:

$$R_{IS} = \frac{\gamma_I}{\gamma_S}\cdot M_2^{IS}\cdot\{\tfrac{1}{2}J(\omega_{1I}-\omega_{1S})-\tfrac{1}{2}J(\omega_{1I}+\omega_{1S})\},$$
$$R_{SI} = \left(\frac{\gamma_S}{\gamma_I}\right)^2\cdot\frac{N_S}{N_I}\cdot R_{IS}, \tag{92}$$
$$R_{II} = M_2^{II}\{J(2\omega_{1I})+\tfrac{5}{3}J(\omega_{0I})+\tfrac{2}{3}J(2\omega_{0I})\}.$$

To obtain $R_{SS}$, the index $I$ has to be replaced by $S$ in the expression for $R_{II}$.

Assuming isotropic thermal motion, the reduced spectral density takes the form

$$J(\omega) = \frac{\tau_c}{1 + (\omega\tau_c)^2}. \tag{93}$$

$M_2^{IS}$, $M_2^{II}$, and $M_2^{SS}$ are the usual Van Vleck second moments. The characteristic inverse time constants are

$$\lambda_\pm = \tfrac{1}{2}\{R_{II} + R_{IO} + R_{SS} + R_{SO}\} \pm \tfrac{1}{2}\{(R_{II} + R_{IO} - R_{SS} - R_{SO})^2 + 4R_{IS}R_{SI}\}^{1/2}. \tag{94}$$

For dilute $S$ spins ($N_S \ll N_I$), the rate constants $R_{SS}$, $R_{IO}$ and the products $R_{IS} \cdot R_{SI}$ are negligibly small compared with $R_{II}$, $R_{SO}$, and with $(R_{II} - R_{SO})^2$, respectively, and therefore the time constants are given by the simplified relations

$$\lambda_+ = R_{II}, \quad \lambda_- = R_{SO} \tag{95}$$

for $R_{SO} > R_{II}$.

Choosing the initial conditions appropriate for a cross-polarization process, viz.

$$M_I(0) = M_{0I}, \quad M_S(0) = 0 \tag{96}$$

the $S$ spin polarization evolves according to the equation

$$M_S(t) = M_{0I} \cdot \frac{R_{SI}}{R_{SO} - R_{II}} (e^{-\lambda_+ \cdot t} - e^{-\lambda_- \cdot t}). \tag{97}$$

Formally, the equation for $M_S(t)$ reveals an analogy with Eq. (37) derived on the basis of the thermodynamic theory for the rigid lattice case, if the correspondences

$$R_{II} = \frac{1}{T_{1\rho I}}, \quad R_{SO} = \frac{1}{T_{CP}} \tag{98}$$

for the rate constants and the expression for the $S$ spin polarization

$$M_{S\infty} = M_{0I} \cdot \frac{R_{SI}}{R_{SO}} \tag{99}$$

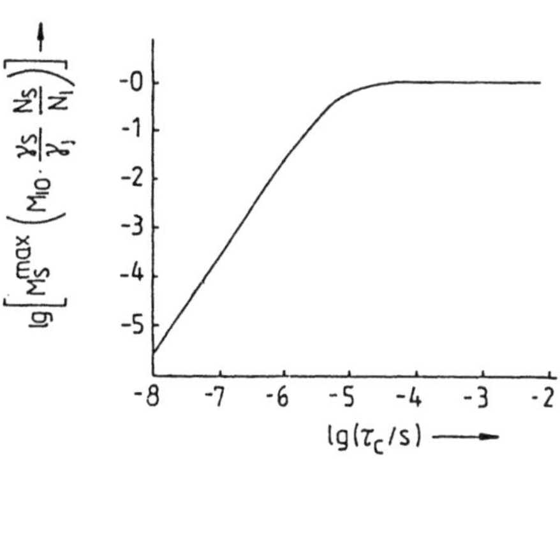


**Fig. 20.** Plot of the logarithm of maximum theoretical $S$ spin magnetization in a standard Hartmann-Hahn CP experiment versus the correlation time $\tau_c$ of an isotropic thermal motion. In the calculation the following parameters were used: $\omega_{0I}/2\pi = 300.13$ MHz, $\omega_{0S}/2\pi = 75.48$ MHz, $\omega_{1I}/2\pi = \omega_{1S}/2\pi = 50$ kHz, $M_{2II} = 6 \times 10^9\,s^{-2}$, $M_{2SS} = 10^9\,s^{-2}$ [186]

are henceforth taken into account. The typical situation for the further analysis is characterized by values for the correlation times, which satisfy the relations

$$\omega_{0S} \gg \tau_c^{-1}, \quad |\omega_{0I} - \omega_{0S}| \gg \tau_c^{-1}$$

such that spectral densities $J(\omega)$ containing $\omega = \omega_{0S}, \omega_{0I}, \omega_{0I} \pm \omega_{0S}$ are negligible in $R_{II}$, $R_{SO}$, and $R_{SI}$ compared to those $J(\omega)$ with $\omega = \omega_{1I}$, $\omega_{1S}$, $\omega_{1I} \pm \omega_{1S}$. Consequently, the rate constants simplify to

$$R_{II} = M_2^{II} \cdot \frac{\tau_c}{1 + (\omega_{1I}\tau_c)^2}$$

$$R_{SO} = M_2^{IS} \cdot \tau_c \cdot \frac{1 + 2(\omega_{1I}\tau_c)^2}{1 + 4(\omega_{1I}\tau_c)^2} \tag{100}$$

$$R_{SI} = \frac{\gamma_S}{\gamma_I} \cdot \frac{N_S}{N_I} \cdot M_2^{IS} \cdot \tau_c \cdot \frac{2(\omega_{1I}\tau_c)^2}{1 + 4(\omega_{1I}^2\tau_c)^2}$$

if at the same time the Hartmann-Hahn matching condition is fulfilled.

### 3.5.3 *Achievable Rare Spin Polarization Under Thermal Motion*

Although the theory for short correlation times is only valid if the correlation frequencies $\tau_c^{-1}$ are larger than about $(M_2^{IS})^{+1/2}$, characterizing the order of magnitude of local magnetic fields, $\tau_c^{-1} \geqslant (M_2^{IS})^{1/2}$, the equation for $M_{S\infty}$ is suitable to estimate the achievable $S$-spin polarization in the limit of extremely long correlation times $(\omega_{1I}\tau_c)^2 \gg 1$, i.e. under conditions which are typical for solids. Taking the ratio $R_{SI}/R_{SO}$ according to Eq. (100), a value

$$M_{S\infty} = M_{0I} \cdot \frac{\gamma_S \cdot N_S}{\gamma_I \cdot N_I} = M_{0S} \cdot \frac{\gamma_I}{\gamma_S} \tag{101}$$

may be derived for the limit $\omega_{1I}\tau_c \to \infty$. This intensity gain in $S$ spin magnetization for cross-polarization of rare spins relative to the static value $M_{0S}$ in normal FID NMR experiments is exactly the same as derived from the thermodynamic theory in Sect. 2.1. It is suggestive to evaluate the maximum of the $S$ spin polarization as a function of the correlation time $\tau_c$. Again, the maximum value is determined by the competing influence of spin-lattice relaxation in the rotating frame and of the polarization transfer. The maximum appears at a time

$$t_m = \frac{\ln(R_{SO}/R_{II})}{R_{SO} - R_{II}}$$

and is characterized by the ratio

$$\frac{M_S^{max}}{M_{0I}} = \frac{R_{SI}}{R_{SO} - R_{II}} [V^{1/(1-V)} - V^{V/(1-V)}].$$

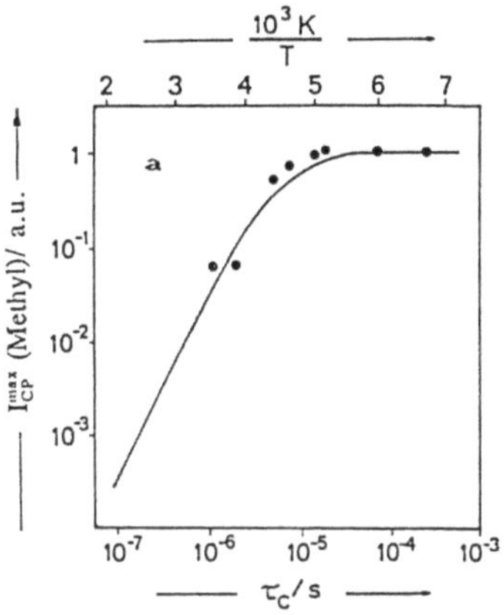


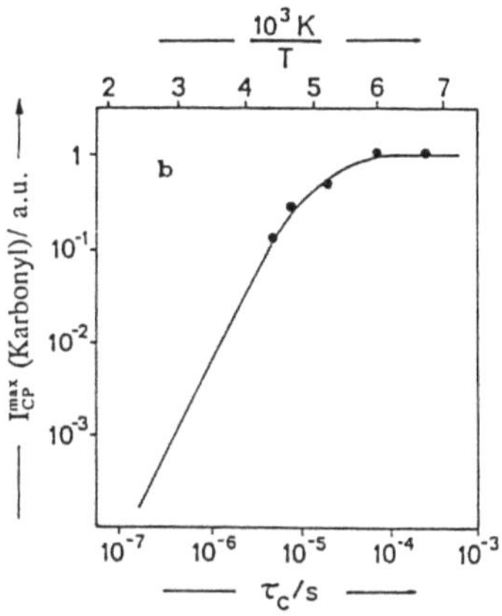


**Fig. 21a, b.** Normalized maximum $^{13}$C–CP NMR line intensities $I_{CP}^{max}$ for the methyl groups **a** and carbonyl groups **b** of acetone sorbed in zeolites of type NaX in dependence on the correlation time $\tau_c$ of the thermal motion. The measurements were run at a $^{13}$C NMR frequency of 75 MHz. The temperature variation of correlation times was measured independently by means of proton spin relaxation [186]. The *lines* connecting the experimental values (·) represent the results of calculations

The relaxation rates $R_{SI}$, $R_{SO}$, and $R_{II}$ are given in eq. (100) and the quantity $V$ is defined by $V = R_{SO}/R_{II}$. A double logarithmic plot of the maximum polarization versus the correlation time $\tau_c$ is shown in Fig. 20, revealing that some critical value $\tau_{c,crit}$ of the correlation time may be defined.

For correlation times longer than $\tau_{c,crit}$ a nearly constant gain of

$$\frac{M_S^\infty}{M_{OS}} \approx \frac{\gamma_I}{\gamma_S}$$

is predicted, while for $\tau_c < \tau_{c,crit}$ the ratio $M_s^{max}/M_{0I}$ diminishes proportional to $(\omega_{1I}\tau_c)^2$. This critical correlation time depends only on $\omega_{1I}$ and the second moments $M_2^{IS}$ and $M_2^{II}$ and possesses typical values at about $10^{-5}$ s. The quadratic decrease appears in the $\tau_c$ range of $10^{-6}$ to $10^{-8}$ s. These theoretical predictions are confirmed by the experiment as shown in Fig. 21.

*Acknowledgement.* The authors are greatly indebted to Professor Bernard C. Gerstein and Dr. M. Pruski (Ames, Iowa) for helpful comments and discussions. They are very grateful to Dr. S. Hediger, Dr. B.H. Meier, and Professor R.R. Ernst (Zürich) for sending the original prints of two figures from their publication [174] and to Mrs. Ursula Seibt and Mrs. Katrin Schandert for their help in preparing the manuscript. Support of the European Commission within the project 7220-EC/004 is gratefully acknowledged.

# 4 References


1. Fyfe CA (1983) Solid-state NMR for chemists CFC Press, Guelph, Canada
2. Hartmann SR, Hahn EL (1962) Phys Rev 128: 2042
3. Jones EP, Hartmann SR (1972) Phys Rev 36: 757
4. Lurie FM, Slichter CP (1964) Phys Rev 133A: 1108
5. Slichter CP, Holton WC (1961) Phys Rev 122: 1701
6. Redfield AG (1955) Phys Rev 98: 1787
7. Pines A, Gibby MG, Waugh JS (1972) J Chem Phys 56: 1176; (1973) J Chem Phys 9: 569; (1972) Chem Phys Lett 15: 373
8. Mansfield P, Grannell PK, Whittaker MAB (1975) Phys Rev 8B: 4149
9. Mehring M (1983) Principles of high resolution NMR in solids. Springer, Berlin Heidelberg New York
10. McArthur DA, Hahn EL, Walstedt RE (1969) Phys Rev 188: 609
11. Demco D, Tegenfeldt J, Waugh JS (1975) Phys Rev B11: 4133
12. Ernst RR, Bodenhausen G, Wokaun A (1990) Principles of nuclear magnetic resonance in one and two dimensions. Clarendon, Oxford
13. Levitt MH, Suter D, Ernst RR (1986) J Chem Phys 84: 4242
14. Freude D, Haase J (1994) Quadrupole effects in solid-state nuclear magnetic resonance, in NMR-Basic Principles and Progress. Springer, Berlin Heidelberg New York 29: 1
15. Engelhardt G, Michel D (1987) High resolution solid state NMR spectroscopy of silicates and zeolites. John Wiley, Chichester
16. Vega S, Pines A (1977) J Chem Phys 66: 5624
17. Vega S (1978) J Chem Phys 68: 5518
18. Wokaun A, Ernst RR (1977) J Chem Phys 67: 1752
19. Pines A (1988) Lectures on pulsed NMR. Proceedings of the 100th School of Physics "Enrico Fermi", North-Holland, Amsterdam, p 43
20. Vega S, Shattuck TW, Pines A (1980) Phys Rev A22: 638
21. Brunner P, Reinhold M, Ernst RR (1980) J Chem Phys 73: 1086
22. Vega S (1981) Phys Rev A23: 3152
23. Woessner DE (1987) Z Phys Chem 152: 51
24. Walter TH, Turner GL, Oldfield E (1988) J Magn Res 76: 106
25. Walter TH, Oldfield E (1989) J Phys Chem 93: 6744
26. Harris RK, Nesbitt GJ (1988) J Magn Res 78: 245
27. Blackwell CS, Patton RL (1984) J Phys Chem 88: 6135
28. Morris HD, Ellis PD (1989) J Am Chem Soc 111: 6045
29. Morris HD, Bank S, Ellis PD (1990) J Phys Chem 94: 3121
30. Kellberg L, Linsten M, Jacobsen HD (1991) Chem Phys Lett 182: 120
31. Rocha J, Liu X, Klinowski J (1991) Chem Phys Lett 182: 531
32. Rocha J, Klinowski J (1991) J Chem Soc, Chem Commun 1121
33. Rocha J, Carr SW, Klinowski J (1991) Chem Phys Lett 187: 401
34. Bryant RG, Ganapathy S, Kennedy SD (1987) J Magn Res 72: 376
35. Edwards JC, Ellis PD (1990) Magn Res Chem 28: S 59
36. Edwards JC, Adams RD, Ellis PD (1990) J Am Chem Soc 112: 8349
37. Vega AJ (1992) J Magn Res 96: 50
38. Vega AJ (1992) Solid State Nucl Magn Res 1: 17
39. Hasse J, Oldfield E (1994) Solid State Nucl Magn Res 3: 171
40. Hasse J, Conradi MS, Oldfield E (1994) J Magn Res A, in press
41. Aue WP, Ruben DJ, Griffin RG (1981) J Magn Res 43: 472
42. Munowitz MG, Griffin RG, Bodenhausen G, Huang TH (1981) J Am Chem Soc 103: 2529
43. Aue WP, Ruben DJ, Griffin RG (1982) J Phys Chem 80: 1729
44. Munowitz MG, Griffin RG (1982) J Chem Phys 76: 2848
45. Munowitz MG, Aue WP, Griffin RG (1982) J Chem Phys 77: 1686
46. Szeverenyi NM, Sullivan MJ, Maciel GE (1982) J Magn Res 47: 462
47. Bax A, Szeverenyi NM, Maciel GE (1983) J Magn Res 51: 400
48. Bax A, Szeverenyi NM, Maciel GE (1983) J Magn Res 52: 147
49. Bax A, Szeverenyi NM, Maciel GE (1983) J Magn Res 55: 494
50. Blümich B, Ziessow D (1983) Mol Phys 48: 955
51. Bronniman CE, Szeverenyi NM, Maciel GE (1983) J Chem Phys 79: 3694

52. Munowitz MG, Griffin RG (1983) J Chem Phys 78: 613
53. Szeverenyi NM, Bax A, Maciel GE (1983) J Am Chem Soc 105: 2579
54. Caravatti P, Neuenschwander P, Ernst RR (1984) in: Müller KA, Kind R, Roos J (eds), Proceedings of the XXII Congress Ampere, Zürich, p. 598
55. DeJong AF, Kentgens APM, Veeman WS (1984) Chem Phys Lett 109: 337
56. Roberts JE, Vega S, Griffin RG (1984) J Am Chem Soc 106: 2506
57. Terao T, Fujii T, Onodera, T, Saika A (1984) Chem Phys Lett 107: 145
58. Harbison GS, Raleigh DP, Herzfeld J Griffin RG (1985) J Magn Res 64: 284
59. Harbison GS, Spiess HW (1986) Chem Phys Lett 124: 128
60. Takegoshi K, McDowell CA (1986) J Chem Phys 84: 2084
61. Terao T, Miura H, Saika A (1986) J Chem Phys 85: 3816
62. Harbison GS, Vogt WD, Spiess HW (1987) J Chem Phys 86: 1206
63. Kentgens APM, DeBoer E, Veeman WS (1987) J Chem Phys 87: 6859
64. Takegoshi K, McDowell CA (1987) J Chem Phys 86: 6077
65. Yang Y, Schuster M, Bümich B, Spiess HW (1987) Chem Phys Lett 139: 239
66. Mass WEJR, Veeman WS (1988) Chem Phys Lett 149: 170
67. Vega AJ (1988) J Am Chem Soc 110: 1049
68. Yang Y, Hagemeyer A, Blümich B, Spiess HW (1988) Chem Phys Lett 150: 1
69. Blümich B, Hagemeyer A (1989) Chem Phys Lett 161: 55
70. Kolbert AC, Raleigh DP, Levitt MH, Griffin RG (1989) J Chem Phys 90: 679
71. Kubo A, Root A, McDowell CA (1990) J Chem Phys 93: 5462
72. Kubo A, McDowell CA (1990) J Chem Phys 92: 7156
73. Tekely P, Montigny F, Canet D, Delpuech JJ (1990) Chem Phys Lett 175: 401
74. Feaux de Lacroix S, Titman JJ, Hagemeyer, Spiess HW (1992) J. Magn Res 97: 435
75. Cheung TTP, Gerstein BC (1981) J Appl Phys 52: 5517
76. Cheung TTP (1981) Phys Rev B23: 1404
77. Cheung TTP (1982) J Chem Phys 76: 1248
78. Carlos WE, Taylor PC (1982) Phys Rev B26: 3605
79. Suter D, Ernst RR (1982) Phys Rev B25: 6038
80. Bronniman CE, Szeverenyi NM, Maciel GE (1983) J Chem Phys 79: 3694
81. Caravatti P, Neuenschwander P, Ernst RR (1984) In: Müller KA, Kind R, Ross J (eds), Proceedings of the XXII Congress Ampere, Zürich, p 598
82. Suter D, Levitt MH, Caravatti P, Ernst RR (1984) ibid., p. 552
83. Suter D, Ernst RR (1985), Phys Rev B32: 5608
84. Henrichs PM, Linder M, Hewitt JM (1986) J Chem Phys 85: 7077
85. Levitt MH, Suter D, Ernst RR (1986) J Chem Phys 84: 4243
86. Takegoshi K, McDowell X (1987) J Chem Phys 86: 6077
87. Suter D, Jarvie TP, Sun B, Pines A (1987) Phys Rev Lett 59: 106
88. VanderHart DL (1987) J Magn Res 72: 13
89. Delpuech JJ, Tekely P, Nicole D, Canet D (1988) Experim Technik Phys (Berlin) 36: 305
90. Kubo A, McDowell CA (1988) J Chem Phys 89: 63
91. Kubo A, McDowell CA (1988) J Chem Soc Faraday Trans 1, 84: 3713
92. Mass WEJR, Veeman WS (1988) Chem Phys Lett 149: 170
93. Tekely P, Brondeau J, Elbayed K, Retournard A, Canet D (1988) J Magn Res 80: 509
94. Tekely P, Canet D, Delpuech JJ (1989) Mol Phys 67: 81
95. Zhang S, Mehring M, Xiaoling W (1990) Chem Phys Lett 166: 92
96. Mass WEJR, VanderHeijden WAC, Veeman WS (1991) J Chem Phys 95: 4698
97. Dixon WT (1981) J. Magn Res 44: 220
98. Aue WP, Ruben DJ, Griffin RG (1982) J Magn Res 46: 354
99. Aue WP, Ruben DJ, Griffin RG (1982) J Chem Phys 80: 1729
100. Dixon WT, Schaefer J, Sefczik MD, Stejskal EO, McKay RA (1982) J Magn Res 49: 341
101. Dixon WT (1982) J Chem Phys 77: 1800
102. Yarim-Agaev Y, Tutunjian PN, Waugh JS (1982), J Magn Res 47: 51
103. Zumbulyadis N (1983), J Magn Res 53: 486
104. Olejniczak ET, Vega S, Griffin RG (1984) J Chem Phys 81: 4804
105. Carduner KR (1987), J Magn Res 72: 173
106. Raleigh DP, Harbison GS, Neiss TG, Roberts JE, Griffin RG (1987) Chem Phys Lett 138: 285
107. Weill G (1987) J Chim Physique 84: 1217
108. Yang Y, Schuster M, Blümich B, Spiess HW (1987) Chem Phys Lett 139: 239
109. Nielsen NC, Bildsoe H, Jakobsen HJ (1988) J Magn Res 80: 149

110. Raleigh DP, Olejniczak ET, Griffin RG (1988) J Chem Phys 89: 1333
111. Yang Y, Hagemeyer A, Blümich B, Spiess HW (1988) Chem Phys Lett 150: 1
112. Engelke F, Maeß B, Fenzke D (1989) J Magn Res 84: 441
113. Feaux de Lacroix S, Titman JJ, Hagemeyer A, Spiess HW (1992) J Magn Res 97: 435
114. Munowitz M (1987) J Magn Res 73: 338
115. Fenzke D, Maeß B, Pfeifer H (1990) J Magn Res 88: 172
116. Jeschke G, Grossmann G (1993) J Magn Res A103: 323
117. Antzukin ON, Song Z, Feng X, Levitt MH (1994) J Chem Phys 100: 130
118. Oas TG, Griffin RG, Levitt MH (1988) J Chem Phys 89: 692
119. Levitt MH, Oas TG, Griffin RG (1988) Israel J Chem 28: 271
120. Raleigh DP, Levitt MH, Griffin MG (1988) Chem Phys Lett 148: 71
121. Levitt MH, Raleigh DP, Creuzet F, Griffin RG (1990) J Chem Phys 92: 6347
122. Schmidt A, Vega S (1992) J Chem Phys 96: 2655
123. Suwelack D, Rothwell WP, Waugh JS (1980) J. Chem Phys 73: 2559
124. Garroway AN, Van der Hart DL, Earl WL (1981) Phil Trans Royal Soc London A 299: 609
125. Van der Hart DL, Earl WL, Garroway AN (1981) J Magn Res 44: 361
126. Naito A, Ganapathy S, Akasaka K, McDowell CA (1983) J Magn Res 54: 226
127. Zumbulyadis N (1983) J Magn Res 53: 486
128. Hemminga MA, der Jager PA, Krüse J, Lamerichs RMJN (1987) J Magn Res 71: 446
129. Xiaoling W, Shanmin Z, Xuewen W (1988) J Magn Res 77: 343
130. Mehring M (1983) Principles of high resolution NMR in solids. Springer, Berlin-Heidelberg New York
131. Gerstein BC, Dybowski CR (1985) Transient techniques in NMR of solids. Academic, Orlando, p 241f
132. Goldman M (1970) Spin temperature and NMR in solids. Oxford University Press, Oxford
133. Levitt MH, Sulter D, Ernst RR (1986) J Chem Phys 84: 4242
134. Jelinski L, Melchior M, Dybowski C, Lichter R (eds) (1987) NMR spectroscopy techniques, Dekker, New York, Chap 6, p 253
135. Zumbulyadis N, O'Reilly JM (1989) J Magn Res 82: 613
136. McArthur DA, Hahn EL, Walstedt RE (1969) Phys Rev 188: 609
137. Demco D, Tegenfeldt J, Wang JS (1975) Phys Rev B11: 4133
138. Zwanzig R (1968) J Chem Phys 33: 1338; Zwanzig R (1964) Physica 30: 1109
139. Mori H (1965) Progr Theor Phys 33: 423
140. Klein-Dowell CH, Mass WEJR, Veemann WS, Buning Werumeus GH, Vankan JMJ (1990) Macromolecules 23: 406
141. Sebald A, Merwin LH, Schaller T, Knöller W (1992) J Magn Res 96: 159
142. Schaller T (1991) Doctoral Thesis, University of Leipzig
143. Jeener J (1968) Adv. Magn Res 3: 205
144. Suter D, Levitt MH, Caravatti P, Ernst RR (1984) Proc. XXII Congress Ampere (Eds. Müller KA, Kind R, Roos J) p. 552, Zürich
145. Zhang S, Wu X, Mehring M (1990) Chem Phys Lett 166: 92
146. Vieth HM, Yannoni CS (1993) Chem Phys Lett 205: 153
147. Zhang S, Steijskal EO, Fornes RE, Wu X (1993) J. Magn Res A 104: 177
148. Stejskal EO, Schaefer J, Waugh JS (1977) J Magn Res 28: 105
149. Dec SF, Wind RA, Maciel GE (1986) J Magn Res 70: 355
150. Wind RA, Dec SF, Lock H, Maciel GE (1988) J. Magn Res 79: 136
151. Stejskal EO, Schaefer J, McKay RA (1984) J Magn Res 57: 471
152. Schaefer J, Stejskal EO, Garbow JR, McKay RA (1984) J. Magn Res 59: 150
153. Pruski M, Dela Rosa L, Gerstein BC (1990) Energy & Fuels 4: 160
154. Sardashti M, Maciel GE (1987) J Magn Res 72: 467
155. Zeigler RC, Wind RA, Maciel GE (1988) J Magn Res 79: 299
156. Barbara TM, Williams EH (1992) J Magn Res 99: 439
157. Wu X, Zilm KW (1993) J Magn Res A 104: 154
158. Peersen OB, Wu X, Kustanovich I, Smith SO (1993) J Magn Res A 104: 334
159. Andrew RE, Bradbury A, Eades RG (1958) Nature 82: 1659
160. Lowe IJ (1959) Phys Rev Lett 2: 285
161. Kessemeier H, Norberg RE (1967) Phys Rev 155: 321
162. Hartmann S, Hahn EL (1962) Phys Rev 128: 2042
163. Engelke F, Kind T, Michel D, Pruski P, Gerstein BC (1991) J. Magn Res 95: 286
164. Engelke F, Habilitationsschrift (1991) Universität Leipzig

165. Mehring M (1983) Principles of high resolution NMR in solids. Springer, Berlin Heidelberg New York, pp. 30–40 and Appendix C
166. Slusher RE, Hahn EL (1968) Phys. Rev 166: 332
167. McArthur DA, Hahn EL, Walstedt RE (1969) Phys Rev 188: 609
168. Müller L, Kumar A, Baumann T, Ernst RR (1974) Phys. Rev Lett 32: 1402
169. Wu X, Zhang S, Wu X (1988) Phys Rev B37: 9827
170. Wu X, Zhang S, Wu X (1988) J. Magn Reson 77: 343
171. Zumbulyadis N (1987) J Chem Phys 86: 1162
172. Wu X, Zilm KW (1993) J Magn Res A 102: 205
173. Gan ZH, Grant DM (1989) Mol Phys 67: 1419
174. Hediger S, Meier BH, Ernst RR (1993) Chem Phys Letters 213: 627
175. Wind RA, Yannoni CS (1986) J Magn Res 68: 373
176. Wind RA, Yannoni CS (1987) J Magn Res 72: 108
177. Abragam A (1961) Principles of nuclear magnetism. Oxford University Press, London
178. Abragam A, Proctor WR (1958) C. R. Acad Sci. 246: 2253
179. Vander Hart DL (1986) J Chem Phys 84: 1196
180. Wind RA, Anthonio FE, Duijvestin MJ, Schmidt J, Trommel J, DeVette GMC (1983) J Magn Res 52: 424.
181. Afeworki M, McKay RA, Schaefer J (1983) Materials Sci and Engin A162: 221
182. Floquet G (1883) Ann. Ecole Normale 12(2): 47; Shirley J (1965) Phys Rev B 138: 979
183. Maricq MM (1982) Phys Rev B 25: 6622
184. Ernst H, Fenzke D, Pfeifer H (1981) Ann. Physik (Leipzig 38: 257
185. Schulze D, Ernst H, Fenzke D, Meiler W, Pfeifer H (1990) J Chem Phys 94: 3499
186. Schulze D, Doctoral Thesis, Univ Leipzig (1990)
187. Vega S, Pines A (1977) J Chem Phys 66: 5624
188. Vega S, Shattuck TW, Pines A (1980) Phys Rev A 22: 638

Mehring M (1983) Principles of high resolution NMR in solids. Springer, Berlin Heidelberg New York, pp 70–80 and references therein

[illegible]

# Solid State NMR Techniques for the Study of Polymer-Polymer Miscibility

**W. S. Veeman[1] and W. E. J. R. Maas[2]**

[1] Physikalische Chemie, Universität-GH-Duisburg, 47048 Duisburg, FRG
[2] Bruker Instruments, Inc. Manning Park, Billerica, MA 01821, USA

**Table of Contents**

Different solid state NMR techniques for the study of the miscibility of polymer blends are reviewed. As an example of such a study experiments on the PMMA/$PVF_2$ blend are reviewed and discussed in a greater detail.

# 1 Introduction

Polymer blends offer a wide range of material properties by varying the composition. Since the physical properties of such blends are strongly influenced by the miscibility and microscopic structure, there is a growing interest in the study of polymer-polymer mixtures as is evident from the increasing number of papers in this field. An important aspect concerning the microstructure of polymer blends, with consequences for the physical properties, forms the heterogeneity with respect to the constituent components or, less precisely, the miscibility. Several techniques are successfully used to determine this heterogeneity: thermodynamical methods, wide- and small- angle X-ray scattering, neutron scattering, light scattering, electron microscopy, mechanical- and dielectric-dynamical methods and nuclear magnetic resonance (NMR).

In many practical instances, the critical property that is used to classify a blend as miscible or not miscible, is the glass transition temperature $T_g$. A blend with one $T_g$, between the $T_g$s of the pure components, is usually classified as miscible and described as single-phase material. This means that if, in such a material, domains consisting of only a single component exist, then they are smaller than a certain dimension. The question, of course, is, how large the size of such single-component domains can be, while the material still macroscopically resembles a well-mixed, single-phase material. Many authors mention that for a single $T_g$ one can expect single-component domains with maximum dimensions of 100–200 Å [1].

Solid state NMR is very well suited to investigating the heterogeneity of polymer blends at the scale below 100–200 Å. As with many other applications of NMR, the enormous freedom and flexibility of the NMR experiment has resulted in many different approaches, each approach employing specific solid state NMR experiments.

At the time of writing, a molecular mixing between blend components has been found with solid state NMR techniques in about sixty polymer blends (Table 1).

This review consists of two parts, in the first part, Sect. 2, we want to discuss the different solid state NMR techniques that have been used to study the miscibility of polymer blends. In the second part, Sect. 3, we review our own work on polyvinylidene ($PVF_2$)/polymethylmethacrylate (PMMA) blends.

Most of the experiments that we will discuss rely heavily on ingenious, but, at the moment, well-known and well-described (solid state) NMR techniques, like Cross-Polarization (CP) [2], Magic Angle Spinning (MAS) [3], Multiple Pulse NMR [4] and Combined Rotation and Multiple Pulse Spectroscopy (CRAMPS) [5], the Nuclear Overhauser Effect [6], Dynamic Nuclear Polarization (DNP) [7] and Rotational Resonance [8]. For a more complete description of these techniques than can be given here, the reader is referred to the specific NMR literature.

**Table 1.** For the following solid polymer blends mixing on a molecular level has been established by one of the NMR methods discussed

The following abbreviations have been used:

| | |
|---|---|
| PS | Polystyrene |
| PVME | Poly(vinylmethylether) |
| PC | Polycarbonate |
| PBMA | Poly(butylmethacrylate) |
| PVPh | Poly(vinylphenol) |
| PMA | Polymethacrylate |
| PEO | Poly(ethyleneoxide) |
| PET | Poly(ethylene terephthalate) |
| PIP | Polyisoprene |
| PCL | Poly(ε-caprolactone) |
| PVA | Poly(vinylalcohol) |
| PS-D | Perdeuterated PS |
| PMMA | Poly(methylmethacrylate) |
| PPO | Poly(phenylene oxide) |
| $PVF_2$ | Poly(vinylidene fluoride) |
| PVAc | Poly(vinylacetate) |
| PVMK | Poly(vinylmethylketone) |
| PA-6 | Polyamide-6 |
| PEMA | Poly(ethylmethacrylate) |
| PVE | Poly(vinylethylene) |
| PAN | Polyacrylonitrile |
| PBT | Poly(butylene terephthalate) |
| PAr | Polyarylate |
| PES | Poly(ethersulfone) |
| PE | Polyethylene |
| PPS | Poly(phenylene sulfide) |
| NECMM | Poly(*N*-ethylcarbazol-3-yl)methylmethacrylate) |
| NECMM | Poly((*N*-ethylcarbazol-3-yl)acrylate) |
| DNBEM | Poly(2-((3,5-dinitrobenzoyl)oxy) ethylmethacrylate |
| Ultem 1000 | Poly(ether-imide) (General Electric) |
| Vectra-A | Copolymer of hydroxybenzoic acid, hydroxynaphthoic acid and poly(ethylene terephthalate) (Hoechst) |

| Blend components | NMR method | Reference |
|---|---|---|
| PS-D/PS-H | MAS proton NMR | 36 |
| PS/PVME | 2D proton spin diffusion | 27 |
| PS/PVME | Selective proton diffusion | 29 |
| PS/PVME | $^{13}C$ NMR | 67 |
| PS/PVME | $^{13}C$ chemical shifts and relaxation | 68 |
| PS/PVME | Intermolecular CP | 69 |
| PS/PVME | NOE | 20, 21 |
| PS/PVME | Intermolecular CP | 70 |
| PS/PVME | Proton spin diffusion | 71 |
| PS-D/PMMA | Intermolecular CP | 72 |
| PS/PC | DNP $^{13}C$ NMR | 35 |
| PS/PPO | Proton relaxation via $^{13}C$ NMR | 33 |
| PS/PPO | $^{31}P$ 2D NMR | 73 |
| PS/PPO | Motions with 2D $^{13}C$ NMR | 74 |
| PS/PBMA | Proton spin diffusion | 75 |
| PS/PS-PB copolymer | Proton spin diffusion via $^{13}C$ NMR | 76 |
| $PMMA/PVF_2$ | Transient NOE | 16 |
| $PMMA/PVPh_2$ | $^{13}C$ chemical shifts; $^{13}C$, $^{1}H$ relaxation | 77 |
| PMMA/PVAc | Proton relaxation | 78 |
| $PMMA/PVF_2$ | Proton relaxation via $^{13}C$ NMR | 51 |
| $PMMA/PVF_2$ | $^{13}C$–$^{19}F$–$^{1}H$ intermolecular CP | 53, 54, 55, 65, 66 |
| $PMMA/PVF_2$ | $^{13}C$ chemical shift | 52 |
| PMMA/PEO | Proton relaxation | 79, 80, 81 |
| PMMA/PVC | Proton relaxation | 82 |
| i-PMMA/s-PMMA | Proton relaxation | 83 |
| s-PMMA/PVC | Proton relaxation | 84 |
| s-PMMA/PVPh | Intermolecular CP | 10 |
| PVPh/PMA | $^{13}C$ chemical shifts; $^{13}C$, $^{1}H$ relaxation | 77 |
| PVPh/PVMK | $^{13}C$ chemical shift and line width | 85 |

**Table 1.** (*Continued*)

| Blend components | NMR method | Reference |
|---|---|---|
| PVPh/PEO | $^{13}C$ chemical shift; proton relaxation | 86 |
| Pa-6/Zn-sulfonated PS | $^{13}C$ NMR; proton relaxation | 87 |
| PA-6/Polyoxadiazole | $^{13}C$ chemical shift | 88, 89 |
| Bisphenol-A PC/PET | Proton spin diffusion | 30 |
| Bisphenol-A PC/PET | $^{13}C$ spin diffusion | 22 |
| Poly(3-octylthiophene)/PPO | Proton relaxation via $^{13}C$ NMR | 90 |
| $PVF_2$/PEMA | Transient NOE | 16 |
| PEO/Phenoxy Resin | Proton relaxation via $^{13}C$ NMR | 91 |
| NECMM/DNBEM | $^{13}C$ chemical shift | 92 |
| PIP/PVE | Proton and $^{13}C$ line widths | 93 |
| Cellulose/PCL | Proton relaxation | 94 |
| Cellulose/PA-6 | Proton relaxation | 94 |
| Cellulose/PAN | Proton relaxation | 94 |
| Cellulose/PVA | Proton relaxation | 94 |
| pDNBEN/pNECMA | $^{13}C$ chemical shift; proton relaxation | 95 |
| pDNBEM/pNECMM | $^{13}C$ chemical shift; proton relaxation | 95 |
| Lignin/Polyurethane | Proton relaxation | 96 |
| PVA/Poly(vinylpyrrolidone) | intermolecular CP | 97 |
| PBT/PAr | Proton relaxation via $^{13}C$ NMR | 98 |
| Polyetherketone/Ultem 1000 | Proton spin diffusion, $^{13}C$ detection | 99 |
| PET/Vectra-A | $^{13}C$ rotational resonance | 23 |
| Polyacetylene/PE | DNP | 34 |
| Poly(2- or 4-methyl PS)/ Poly(2,6-dimethyl PPO) | Proton spin diffusion | 100 |
| Polybenzimidazole/Polyimide | Proton relaxation | 101 |
| PES/Polyimide | Proton relaxation | 102 |
| PPS/PES | Proton relaxation | 103 |

# 2 Solid State NMR and Polymer Blends

## 2.1 *The Dipolar Interaction*

All NMR studies of polymer-polymer miscibility have in common that some NMR parameter of one component is measured that is sensitive to the near-presence of another component. Although it does happen that the chemical shift of a nuclear spin in component *A* is affected by the presence of units of component *B*, most studies rely in some way on the magnetic interaction between neighbouring spin dipoles, the dipolar interaction. For two spins with angular momentum operators $I_1$ and $I_2$ and internuclear vector $r_{12}$, the dipolar interaction can be written in the form:

$$H_D = \frac{\gamma^2\hbar^2}{r_{12}^3}\left[I_1 \cdot I_2 - 3\frac{(I_1 \cdot r_{12})(I_2 \cdot r_{12})}{r_{12}^2}\right] \tag{1}$$

or as:

$$H_D = \frac{\gamma^2\hbar^2}{r_{12}^3}[A + B + C + D + E + F] \tag{2}$$

where

$$A = I_{1z}I_{2z}(1 - 3\cos^2\theta)$$
$$B = -\tfrac{1}{4}(I_1^+ I_2^- + I_2^+ I_1^-)(1 - 3\cos^2\theta)$$
$$C = -\tfrac{3}{2}(I_1^+ I_{2z} + I_{1z}I_2^+)\sin\theta\cos\theta\, e^{-i\phi}$$
$$D = -\tfrac{3}{2}(I_1^- I_{2z} + I_{1z}I_2^-)\sin\theta\cos\theta\, e^{i\phi}$$
$$E = -\tfrac{3}{4}I_1^+ I_2^+ \sin^2\theta\, e^{-2i\phi}$$
$$F = -\tfrac{3}{4}I_1^- I_2^- \sin^2\theta\, e^{2i\phi}$$

and the angles $\theta$ and $\phi$ define the internuclear vector in an axis system with the external magnetic field along the z axis.

This dipolar interaction is the cause that, in general, a spin feels a local field due to the presence of other spins. Due to the fact that the nuclear magnetic moments are small and that the magnitude of this local field decreases sharply ($r^{-3}$) with the internuclear distance, the presence of a nucleus can be felt only over short distances, ranging from a few Å to distances on the order of 100 Å, depending on the technique in question. The NMR methods which we will discuss, are therefore excellently suited to studying the miscibility of polymer blends on a molecular scale.

The various NMR techniques can be distinguished in different ways. Here we make the division into really short-range techniques (up to about 6 Å) and techniques that are suitable for studying the miscibility of polymers over distances greater than 6 Å, up to distances of about 100 Å.

## 2.2 Short Range Techniques

At least five techniques that have been reported for the study of miscibility of polymer blends fall into the short range category. They are based on the use of the chemical shift as a local probe, or employ one of the following (solid state) NMR techniques: cross-polarization, the nuclear Overhauser effect, two-dimensional heteronuclear correlation NMR or rotor-driven $^{13}C$ spin diffusion.

### 2.2.1 Short-Range Techniques Based on Carbon-13 Chemical Shifts

The chemical shift of a nucleus is sensitive to the local electron density. When, in a polymer blend, an interaction occurs between the different components, it may affect the electron density locally near an NMR nucleus and change the chemical shift. Since the spectral resolution in a solid state NMR spectrum is rather low, to observe such a change in chemical shift two conditions have to be fulfilled. First, the interaction between the components of the blend has to be relatively strong. Direct detection of an intermolecular interaction in polymer blends has been reported only for cases were hydrogen bonding or charge transfer occurs. The second condition is that the spectrum shows high resolution. A resolution high enough to detect such changes in chemical shift has only been obtained with carbon-13 NMR where the line broadening due to dipolar couplings with protons has been removed by dipolar decoupling (DD) and the line broadening due to the chemical shift anisotropy by magic angle spinning (MAS). Although changes in chemical shift in a solid state proton spectrum obtained under high resolution conditions with multiple pulse decoupling and MAS (CRAMPS) have been reported for a complex between a polymer and a monomer [9], as far as we know, no such studies have been reported for polymer blends.

### 2.2.2 *Intermolecular Cross-Polarization (CP)*

CP makes use of the fact that when two different nuclear spin systems $I$ and $S$, coupled via dipolar interactions, are irradiated with two radiofrequency (RF) fields, one approximately resonant for the $I$ spins and the other approximately resonant for the $S$ spins, for a certain combination of the rf field strengths (the Hartmann-Hahn condition) magnetization can be transferred from one spin system to the other. In most CP experiments [2] magnetization is transferred from abundant proton spins to rare carbon-13 spins to enhance the carbon-13 signal. This results in relatively strong and highly-resolved carbon-13 spectra. However, since the CP transfer is mediated by the short range dipolar interactions of Eq. (1) or (2), CP can also be used to detect whether $I$ and $S$ spins are physically near each other.

Many studies of polymer miscibility make use of this fact. Let us assume that in a binary blend, one of the components, $A$, has only spins $I$ and both components $A$ and $B$ have spins $S$. Further we assume that in the $S$ spin spectrum, obtained with the highest possible resolution, at least one $S$ spin NMR line of component $B$ at frequency $\nu_B$ does not overlap with any of the $S$ spin NMR lines of component $A$. When in such a situation a CP transfer of magnetization from $I$ spins in component $A$ to $S$ spins in $A$ and $B$ results in intensity at frequency $\nu_B$, then this proves that the $S$ spins responsible for the intensity at $\nu_B$ are only a few Å separated from an $I$ spin. Situations are also found where only component $B$ has $S$ spins.

The distinction between the different studies that have been reported with this technique, lies in the choice of nuclear spins $I$ and $S$. A popular approach is to perdeuterate one of the components in the blend, say B. Both components contain carbon-13 spins and $^{1}H-^{13}C$ CP from protons in A to carbons in $B$ can establish whether the two components are mixed on a molecular scale. Also direct cross-polarization between protons and deuterons can be employed [10]. A disadvantage of this technique is that the miscibility between two polymers may be affected when one of the components has to be deuterated. In fact, it has been reported that even polymer isotope mixtures are not simple mixtures [11, 12].

In Sect. 3 another example of this technique is discussed at great length; here the CP transfer is between fluorine-19 spins and carbon-13 spins.

An important aspect of the CP technique for the study of blends is to obtain distance information from such experiments. For a perfectly tuned CP process between $N_I$ abundant $I$ spins (spin quantum number $I$) and $N_S$ dilute $S$ spins (spin quantum number $S$) the signal intensity of the $S$ spins $I_S(t)$ as a function of the $I \Rightarrow S$ CP transfer time $t$ can be expessed as [13]:

$$I_S(t) = \frac{I_S(\infty)}{a_+ - a_-}\left[\exp\left(\frac{-a_- t}{T_{IS}}\right) - \exp\left(\frac{-a_+ t}{T_{IS}}\right)\right] \tag{3}$$

where

$$a_\pm = a_0[1 \pm \sqrt{1 - b/a_0^2}]$$

$$a_0 = \frac{1}{2}\left[1 + \varepsilon' + \frac{T_{IS}}{T_{1\rho}^I} + \frac{T_{IS}}{T_{1\rho}^S}\right]$$

$$b = \frac{T_{IS}}{T_{1\rho}^I}\left(1 + \frac{T_{IS}}{T_{1\rho}^S}\right) + \varepsilon'\frac{T_{IS}}{T_{1\rho}^S}$$

$$\varepsilon' = \frac{N_S S(S+1)}{N_I I(I+1)}$$

$T_{IS}$ is the characteristic time for the CP transfer, $T_{1\rho}$ is the spin-lattice relaxation time in the rotating frame for the $I$ or $S$ spins. $I_S(\infty)$ is the maximum $S$ spin magnetization available through CP in the absence of relaxation processes. Demco et al. [14] have approximated the CP rate $T_{IS}^{-1}$ as:

$$T_{IS}^{-1} = \tfrac{1}{2} M_{2,SI} J_X \tag{4}$$

where

$$M_{2,SI} = \frac{\langle \Delta\omega^2 \rangle}{4\pi^2}$$

is the van Vleck second moment of the $S$ spin resonance line determined by the dipolar coupling of the $S$ spins with $I$ spins and $J_X$ is the cross-polarization spectral density function of the $I$ spins. The second moment expresses the static dipolar interaction between $S$ and $I$ spins, while the spectral density function is determined by fluctuations in the $I$ spin system due to spin flip-flop transitions induced by the $B$ term of Eq. (2). The spectral density function can be approximated by assuming a Gaussian shape for the correlation function [14], leading to the following expression for $J_X$:

$$J_X = \tfrac{1}{2}\sqrt{\pi}\,\tau_c \tag{5}$$

where $\tau_c$ is the correlation time which is on the order of the $I$ spin $T_2$ and can be estimated from the $I$ spin line width. The second moment can be described, according to Slichter [15], as:

$$\langle \Delta\omega^2 \rangle = \frac{1}{3}\gamma_I^2\gamma_S^2\hbar^2\left(\frac{\mu_0}{4\pi}\right)^2 I(I+1)\sum_{I,S}(1-3\cos^2\theta_{IS})^2 r_{IS}^{-6} \tag{6}$$

in which $r_{IS}$ is the distance between $I$ and $S$ spins and $\theta_{IS}$ the angle between the internuclear vector and the external field direction. For a given $S$ spin the second moment contains contributions from all $I$ spins that are dipolar coupled to this $S$ spin. Because of the $r^{-6}$ dependence of the second moment, the summation in Eq. (6) can be limited to nearest neighbors. For a polycrystalline solid with the powder average of $(1-3\cos^2\theta_{IS})$:

$$\langle \Delta\omega^2 \rangle = \frac{4}{15}\gamma_I^2\gamma_S^2\hbar^2\left(\frac{\mu_0}{4\pi}\right)^2 I(I+1)\sum_{I,S} r_{IS}^{-6} \tag{7}$$

From a CP experiment where the $S$ spin signal intensity is measured as a function of the contact time $t$, $T_{IS}$ can be determined from the fit of Eq. (3) to the experimental data. For an accurate determination of $T_{IS}$ the carbon-13 and proton $T_{1\rho}$ have to be measured in separate experiments. Once the CP time $T_{IS}$ and the correlation time $\tau_c$ are known, the term:

$$\sum_{I,S} r_{IS}^{-6} \tag{8}$$

can be calculated with Eqs. (4), (5) and (7). Especially for the situation of a polymer blend where we cross-polarize from $I$ spins in one component to $S$ spins in the other component, one can assume that one $I-S$ distance is shorter than all others. Due to the $r^{-6}$ dependence it then seems reasonable to approximate the sum in Eq. (8) by:

$$\sum_{I,S} r_{IS}^{-6} \approx \langle r \rangle_{IS}^{-6} \tag{9}$$

where $\langle r \rangle_{IS}$ is the average nearest $I-S$ distance.

By combining Eqs. (4), (5), (7) and (9) and taking $I = 1/2$, we find for protons and carbon-13 spins:

$$\langle r \rangle_{IS} = \sqrt[6]{\frac{\pi^{-1.5}}{80}\gamma_I^2\gamma_S^2\hbar^2\left(\frac{\mu_0}{4\pi}\right)^2 \tau_c T_{IS}} = \sqrt[6]{7.8 * 10^{-53}\tau_c T_{IS}} \tag{10}$$

In practical situations, the range of $T_{IS}$ values that can be measured is limited by the length of the proton and carbon-13 $T_{1\rho}$ values. We can therefore estimate the longest $\langle r \rangle_{IS}$ that one can hope to find from a CP experiment. When we assume that in most cases $T_{IS}$ values longer than about 30 ms cannot be determined, we find for the $^1H$-$^{13}C$ distance in a rigid proton system with $\tau_c = 1$ ms:

$$\langle r \rangle_{max} \approx 3.6\,\text{Å} \tag{11}$$

### 2.2.3 *Nuclear Overhauser Effect*

Two-dimensional nuclear Overhauser effect spectroscopy (2D-NOESY) is an important technique for structural and conformational analysis of biomacromolecules in solution. The nuclear Overhauser effect (NOE) also plays an important role in a number of blend NMR studies. An early example is the work of Douglass and McBrierty [16] who exploited the transient NOE in blends of $PVF_2$ and PMMA and of $PVF_2$ and polyethylmethacrylate (PEMA). Here the effects of spin diffusion show that these experiments should be classified as belonging to the long range category. 2D-NOESY studies of blends in concentrated solutions have been performed by Kögler and Mirau [17], by Mirau et al. [18] and by Kelts et al. [19]. White and Mirau [20] used the NOE technique in a one-dimensional fashion to study the miscibility of deuterated polystyrene (PS) and poly(vinyl methyl ether) (PVME) in the solid phase. They were able to show that the motion of the methyl group protons in PVME modulate the dipolar local field at the PS phenyl ring carbons in blends cast from toluene, in contrast to blends cast from chloroform.

### 2.2.4 *Two-Dimensional $^{13}C$-$^{1}H$ Correlation NMR*

In order to correlate a high resolution carbon-13 spectrum with a high resolution proton spectrum for solids in a two-dimensional fashion, three conditions have to be fulfilled: a) during the evolution period the proton-proton, proton-carbon dipolar couplings and the proton chemical shift anisotropy have to be eliminated, b) during the mixing period coherence has to be transferred from the proton spins to the carbons and c) during the detection period a high-resolution carbon-13 spectrum has to be acquired. White and Mirau [21] have shown that two-dimensional $^{13}C$-$^{1}H$ correlated spectra can be obtained from a polymer blend when, during the evolution time, homo- and hetero-nuclear dipolar couplings are eliminated by BLEW-12 and BB-12 pulse sequences, and when coherence transfer is induced by a WIM-24 mixing cycle. The carbon-13 spectrum is obtained under proton decoupling conditions, while MAS eliminates all line boadenings due to proton and carbon chemical shift anisotropies. Since the distance determining step basically is a one-step process mediated by the dipolar coupling between protons and carbons, this technique should be listed under the techniques that can provide information about short range effects. With this technique a direct interaction between a hydroxyl hydrogen of poly(vinyl phenol) and a carbonyl carbon of poly(methyl acrylate) could be detected.

### 2.2.5 *Rotational Resonance*

Although rotational resonance or rotor-driven $^{13}C$ spin-diffusion is an example of the spin-diffusion methods discussed in the next section, it is mentioned here since it seems that the distance between molecules that can be bridged by this technique falls in the range covered in this section. Due to the $B$ term of the dipolar interaction two spins $I_1$ and $I_2$ can undergo flip-flop transitions. In the two-spin system with spin states (for both spins with $I = 1/2$) $|\alpha\alpha\rangle$, $|\alpha\beta\rangle$, $|\beta\alpha\rangle$ and $|\beta\beta\rangle$, such transitions correspond to transitions between $|\alpha\beta\rangle$ and $|\beta\alpha\rangle$. When the two spins $I_1$ and $I_2$ are both carbon-13 spins under $^1H$ decoupling conditions, the probability for such transitions is negligible, unless the coupling $B$ term is at least of the same order of magnitude as $\Delta\delta$, the difference in chemical shift (in energy units) of the two spins. Therefore, normally spin diffusion between $^{13}C$ spins is very slow and not so attractive for the study of the miscibility of polymer blends. For carbon-enriched systems the average dipolar interaction is drastically increased and carbon-carbon spin diffusion times are reduced by two orders of magnitude. Spin exchange between two carbons of two components in a blend can then be used to obtain information about specific nearest-neighbor interactions. We will not discuss this technique, a good overview has been given by Linder et al. [22]. Here we will discuss another technique to enhance the carbon-carbon flip-flop rate, rotor-driven spin diffusion or rotational resonance [8]. To describe how this rotor-driven spin-diffusion works, we use a simplified description that employs the equivalence of $^{13}C$ spin diffusion and transitions between the $|\alpha\beta\rangle$ and $|\beta\alpha\rangle$ spin states of the four-level two-spin system mentioned above. In the case that the dipolar interaction between the spins is much smaller than the chemical shift difference $\Delta\delta$, the two spin states $|\alpha\beta\rangle$ and $|\beta\alpha\rangle$ have an energy difference of $\Delta\delta$. Magic angle spinning with frequency $\omega_R$ renders the dipolar interaction time-dependent with frequencies $\omega_R$ and $2\omega_R$ [15]. In other words, magic angle spinning generates in the Hamiltonian a time-dependent spin operator $(I_{1+}I_{2-} + I_{1-}I_{2+})$ with frequencies $\omega_R$ and $2\omega_R$. When one of these frequencies, say $\omega_R$, equals the splitting between $|\alpha\beta\rangle$ and $|\beta\alpha\rangle$:

$$\hbar\omega_R = \Delta\delta,$$

transitions between the two spin states are induced by the time-dependent $B$ term of the dipolar interaction.

This simplified description shows that rotational resonance can occur only for two rotor frequencies $\omega_R = \Delta\delta/\hbar$ and $\omega_R = \hbar\Delta\delta/2$. A complete description, taking into account that the chemical shift can also be anisotropic, makes it clear that rotational resonances of higher order can also occur. The rotor-driven $^{13}C$ spin diffusion experiment can be employed to study the miscibility in blends by using a standard two-dimensional exchange experiment [23]. When the magic angle spinning frequency is tuned to the frequency difference of two carbons $C_A$ and $C_B$, where $C_A$ is in component $A$ and $C_B$ in component $B$, then the occurrence of cross-peaks between the diagonal lines of $C_A$ and $C_B$ implies miscibility on a molecular scale between polymer $A$ and polymer $B$. So far only carbon-13 enriched samples have been used [23].

## 2.3 Long Range Techniques

### 2.3.1 Spin Diffusion

All long range techniques rely in some way on the spin diffusion process. Spin diffusion is, like rotational resonance, a consequence of the $B$ term of the dipolar interaction of Eq. (2). For a large network of dipolarly coupled spins, the effect of successive energy-conserving spin flip-flops can be described as a diffusion process where a local disturbance of the spin system (a non-equilibrium value of the spin magnetization) is dissipated to sites spatially removed from the original disturbance. A radial spin diffusion process with an isotropic diffusion constant $D$ can be described by the diffusion equation [24]:

$$\frac{\partial m(r,t)}{\partial t} = D\frac{\partial^2 m(r,t)}{\partial r^2} \tag{12}$$

where $m(r,t)$ represents the magnetization at time $t$ at position $r$. For spherical diffusion from a point source at $r = 0$, the solution of Eq. (12) is [24]:

$$m(r,t) = \frac{m_0}{8(\pi Dt)^{3/2}} \exp\left(\frac{-r^2}{4Dt}\right) \tag{13}$$

where $m_0$ is the total magnetization. The fraction of the magnetization in the volume $4\pi r^2\, dr$ between two spheres with radii $r$ and $r + dr$, respectively, at time $t$ is:

$$4\pi r^2 dr \frac{m(r,t)}{m_0}$$

The mean square distance $\langle r^2 \rangle$ this magnetization has moved from the point source in a time $t$ is given by:

$$\langle r^2 \rangle = \int_0^\infty 4\pi r^4 \frac{m(r,t)}{m_0} dr = 6Dt \tag{14}$$

To simplify the problem, we assume that after a disturbance at some point, spin diffusion during a time $t$ can equalize the magnetization in a spherical domain with radius $\sqrt{6Dt}$. Some workers have assumed a lamellar morphology for polymer blends and spin diffusion must then be described as an one-dimensional process. Since the root-mean-square distance travelled for such a linear process differs only by a small factor from the three-dimensional case, we will not worry here about the best model and simply assume that the three-dimensional model describes spin diffusion in polymer blends. For a recent and more extensive discussion of spin diffusion see Clauss et al. [25].

As Eq. (14) shows, the process of spin diffusion can provide information about domain sizes in polymer blends. Although spin diffusion may occur in every spin system, it will be clear that the diffusion rate can be relatively high only for dense spin systems with like spins. For carbon-13 spins, for example, the average dipolar interaction is very small due to the large, average internuclear distances. For such spins the spin diffusion is slow, especially because in general the dipolar interaction is much smaller than the chemical shift differences. Although one can think of several experiments that eliminate this chemical shift difference, thereby increasing the spin diffusion rate by more than one order of magnitude (e.g. spin locking [26], rotational resonance, see Sect. 2.2.5), we will not consider them here. We will limit ourselves in the discussion of this section to spin diffusion among proton spins.

We can distinguish direct spin diffusion experiments where the rate of diffusion during a certain time is measured, and indirect experiments where some other parameter is measured whose magnitude is determined by spin diffusion. A direct experiment always consists of two parts. In the first part the spins are prepared in such a way that different domains have different magnetizations. In the second part, spin diffusion restores equilibrium and the rate of this relaxation process is measured and related to an equation like Eq. (14). We will first discuss the direct experiments, the different ways the magnetization gradient can be generated and how spin diffusion is manifested.

### *2.3.2 Direct Spin Diffusion Experiments*

For a direct proton spin diffusion experiment on a binary polymer blend $A/B$, proton spins in $A$ have to differ in some respect from spins in $B$, for example, because of a different chemical shift, or because of a difference in relaxation behavior. In case of a different chemical shift two possible ways are open: a two-dimensional experiment or selective excitation. An early example of the study of the miscibility in polymer blends forms the two-dimensional proton experiment by Caravatti et al. [27] In this Jeener exchange experiment [28] on PS/PVME blends the aromatic protons of PS and the aliphatic protons in PVME are distinguished in the evolution and detection period by eliminating the line broadening due to dipolar interactions and chemical shift anisotropy in a CRAMPS experiment, while during the mixing period spin diffusion takes care of the exchange of magnetization between PS and PVME protons. The occurrence of such spin diffusion, and the proof that the two components are mixed on a molecular scale, is shown by the presence of cross-peaks in the two-dimensional spectrum. The same authors [29] showed for the same blend that the spectral resolution also allowed selective excitation experiments, from which could be confirmed that the PS/PVME blend can be described as a three-phase system.

In these experiments magnetization differences are generated among the proton spins and also the proton spins are detected. An obvious extension of this experiment is to monitor the progression of proton spin diffusion via cross-polarization to the carbon spins. The much higher resolution in the carbon-13 spectrum can be exploited to improve the selectivity [25, 30].

A much used technique to study the proton spin diffusion is based on differences in mobility of the two components forming a binary blend. In the original approach of Goldman and Shen [31] a magnetization gradient can be achieved based on the different dipolar strengths of the two components due to molecular motions. Since the dipolar interaction is orientation dependent, a relatively fast molecular motion may partly or completely average out the dipolar interaction. This means that, for a blend where the two components exhibit significantly different molecular mobilities, the transverse magnetization of one component decays much faster than of the other. A magnetization gradient is then created which spin diffusion tries to eliminate. Of course, the decay of the transverse magnetization described by a $T_2$ relaxation time, is just one of the relaxation processes and differences between other relaxation parameters can also be employed to generate magnetization gradients.

### 2.3.3 *Indirect Spin Diffusion Effects*

The most common technique for the study of miscibility in polymer blends uses the fact that when the two separate components have different proton relaxation times, mostly $T_1$ and $T_{1\rho}$ are used, spin diffusion between protons in the blend may or may not equalize the two values depending on the size of the domains. The procedure is simple, measure the appropriate relaxation time and compare with the component values. There are two limiting possibilities: the blend relaxation process can be described by a single exponential or by two exponentials. In the first case one can assume that the domains are small enough for spin diffusion to average the values from the two components, in the latter case the domains are too large for spin diffusion to equalize the individual relaxation times. When the relaxation times of component $A$ and $B$ are $T_A$ and $T_B$, the blend relaxation time $T_K$ averaged by spin diffusion can then be written as

$$T_k^{-1} = p_A T_{kA}^{-1} + p_B T_{kB}^{-1} \tag{15}$$

where $p_A$ and $p_B$ represent the relative proton signal intensities for component $A$ and $B$ and $T_k$ is $T_1$ or $T_{1\rho}$.

From our Eq. (14), we can immediately estimate the limiting size $L_D$ of the domains by realizing that spin diffusion can proceed during a time of the order of the appropriate relaxation time $T_k$:

$$L_D = \sqrt{\langle r^2 \rangle} = \sqrt{6DT_k} \tag{16}$$

The diffusion coefficient $D$ depends on the density of the proton spins and the proton linewidth. It is generally assumed that $D$ scales with the proton density as the 1/3 power [32]. In many systems $D$ is on the order of $10^{-11}$ to $10^{-12}$ cm²/s, for polyethylene for instance $D = 6.2 \times 10^{-12}$ cm²/s [32]. When we take average values of $D = 5 \times 10^{-12}$ cm²/s and of the proton $T_{1\rho} = 10$ ms and of the proton

$T_1 = 1\text{s}$, we find for $L_D$ in case of $T_{1\rho}$ experiments (remember that for spin diffusion in the rotating frame $D$ is scaled by a factor 1/2):

$$L_D(T_{1\rho}) \approx 40\,\text{Å}$$

and for $T_1$ experiments:

$$L_D(T_1) \approx 500\,\text{Å}$$

With these values one can make a quick estimate of domain sizes in a blend by measuring proton $T_1$ and $T_{1\rho}$ values. Assuming that the component relaxation times are sufficiently different, one has the following possibilities:

a) the blend has one proton $T_1$ and one proton $T_{1\rho}$ $\Rightarrow$ domains, when present, are smaller than $\approx 50$ Å, molecular mixing;
b) the blend has one proton $T_{1\rho}$ and two $T_1$s $\Rightarrow$ average domain sizes are between 50 and 500 Å;
c) blend has two values for each type of spin-lattice relaxation time $T_1$ and $T_{1\rho}$ $\Rightarrow$ average domain size larger than 500 Å.

So far we have discussed only proton spin-lattice relaxation, further information about miscibility can also be obtained from carbon-13 relaxation parameters. However, because carbon-13 spin diffusion, as discussed above, is a very slow process in general, the information from such experiments usually is not as clear as from proton measurements. Therefore we will not discuss such experiments here.

A significant improvement of the proton spin-lattice relaxation experiment for miscibility determinations can be obtained when the actual detection of the proton signal is made indirectly via the carbon-13 spins [33]. For that purpose, one transfers, at the end of the spin-lattice relaxation measurement, proton magnetization to the carbon spins via cross-polarization. When one component of the blend has at least one carbon-13 NMR line that does not overlap with the carbon lines of the other component, the proton relaxation of that component can then be separately measured. In that way there is no need to decompose proton spin-lattice relaxation curves in order to see whether there is one relaxation time or two.

### *2.3.4 Dynamic Nuclear Polarization (DNP)*

So far we have discussed NMR techniques that are suitable to investigate miscible blends or, in other words, blends where the total interface area between the components is very large. For immiscible blends, all the techniques we have discussed so far, do not work because the interface area is too small and the number of spins in these interface areas is not large enough to be detected. NMR basically is a bulk technique. With DNP it is possible to enhance the signal from nuclei in interfaces between polymer components of which one has unpaired electrons. In this way two advantages are obtained, first, since the DNP mechanism relies again on the dipolar interaction, this time between electron spins and nuclear spins, the spins that are influenced by the DNP process have to be close to the interface, and secondly, the detection limit of such spins is appreciably lowered. This technique has been applied to polyacetylene/polyethylene [34] and to polystyrene/polycarbonate blends [35]. For a discussion of the DNP effect and of the application to blends we refer to these publications. Here it suffices to say that the experiment is rather involved while, in addition to the normal solid state carbon-13 NMR experiment with magic angle spinning, one also has to saturate the electron spin resonance (ESR) line of the unpaired electrons. The success of the DNP method is further crucially determined by the ESR line width and other parameters. For the cases where it works, it is an unique method since one can observe spins in areas which usually are not accessible to NMR observation but of high interest for polymer research.

The DNP process can be used for direct electron-carbon-13 transfer or for electron-proton transfer, followed by proton spin diffusion and proton-carbon-13 transfer via CP. The poblem with the first process seems to be that carbons with sufficiently strong coupling with the electrons to show DNP enhancements, also have, because of this interaction, broadened NMR lines [35].

## *2.4 Other NMR Methods*

In the two foregoing sections, we discussed the main methods to study miscibility of polymer blends by solid state NMR. Other techniques, however, are certainly worth mentioning. VanderHart et al. [36] have shown that in a mixture where one component is deuterated, the dilute protons in the 98–99% deuterated component show a MAS spectrum with a centerband and spinning sidebands on top of the broad proton spectrum from the protonated component, when the mixing is not intimate. When the environment of the deuterated units contains protons from the protonated component, in the case of intimate mixing, this results in a significant weakening of or complete disappearance of the spinning sideband pattern.

$^{129}$Xe NMR has also been used to study polymer blends [37, 38]. It is well-known that the chemical shift of Xe is very sensitive to its environment and when the Xe atoms reside in the free volume of an immiscible polymer blend, it may show more than one line at temperatures low enough that no rapid exchange between the different adsorption sites takes place [39].

The last technique we will mention is that of NMR imaging. The problem with this technique is the low inherent sensitivity of NMR and consequently, the poor spatial resolution. However, the combination of imaging with further discrimination with the help of NMR parameters such as relaxation times or chemical shift, makes it an attractive technique and some images of heterogeneous materials have been published [40, 41].

### 2.5 Conclusion

Several ingenious solid state NMR techniques have been used to investigate the miscibility of polymer blends in the solid phase. Table 1 shows all polymer blends for which an intermolecular interaction between the components has been detected with one of the methods discussed above.

## 3 Intermolecular Cross-Polarization and -Depolarization of $PVF_2$/PMMA Blends

### 3.1 Introduction

It is well established that poly(methyl methacrylate) (PMMA) and poly-(vinylidene fluoride) ($PVF_2$) are completely miscible [42, 43] over the entire composition range in the molten state below the lower critical solution temperature (LCST) at $\approx 350\,^{\circ}C$ [44]. The Flory-Huggins interaction parameter $\chi$ has been reported to be about $-0.3$ [43, 45, 46], which is one of the most negative values reported in the literature. This indicates strong interactions between the components of the blend, which lead to the observed miscibility.

Solid $PVF_2$ is partially crystalline, it can exist in a variety of crystalline forms, which are referred to as the $\alpha, \beta, \gamma$ and $\alpha_P$ modifications [47]. In a solid blend, this crystallization interferes, at higher $PVF_2$ content, with the formation of a miscible, single, amorphous phase. Depending on the method of blend preparation, the amount and also the crystal form of the crystallites can be influenced. X-ray and dielectric relaxation studies by Hahn et al. [48] revealed that, when $PVF_2$ crystallizes in the blend, a triphasic system is formed, consisting of $PVF_2$ crystallites, an amorphous $PVF_2$ phase, and a miscible amorphous PMMA/$PVF_2$ phase.

The PMMA/$PVF_2$ blend has also been studied using various NMR techniques [16, 49, 50, 51, 52]. The results obtained in these studies were very dependent on blend preparation. Douglass and McBrierty [16] measured cross relaxation effects between $^1H$ and $^{19}F$ by transient Overhauser effects on blends that contained a crystalline $PVF_2$ phase. They concluded that at least 20% of the amorphous $PVF_2$ in the PMMA/$PVF_2$ 40/60 blend "sees" PMMA molecules at nearest neighbor distances.

Ward and Lin [49] observed intensity attenuations of the PMMA signals due to strong intermolecular dipolar interactions between the fluorines in $PVF_2$ and the carbons in PMMA [49, 50], indicating intimate mixing.

In the following sections, we will discuss three solid state NMR techniques which provide detailed information on the miscibility behavior of PMMA and $PVF_2$ on a molecular scale.

The techniques used to acquire information on molecular miscibility, rely on magnetization transfer via CP between three nuclei of $PVF_2$ and PMMA.

In the following sections we will discuss successively fluorine-carbon CP, fluorine-proton-carbon double-CP and proton-fluorine cross-depolarization, followed by proton-carbon CP. These experiments require that three nuclear spins can be irradiated: $^{1}H$, $^{19}F$ and $^{13}C$, while only the last nucleus is observed. The spectrometer configuration is described in detail elsewhere [53].

## 3.2 $^{19}F-^{13}C$ Cross-Polarization

In the fluorine-carbon cross CP experiment, magnetization is transferred from fluorines of $PVF_2$ to carbons of PMMA (see Fig. 1a). As discussed in Sect. 2.2.2 CP is effective only over maximum distances of about 4 Å. The occurrence of PMMA resonances in a $^{19}F-^{13}C$ CP spectrum is therefore a direct indication of molecular miscibility in the blend. In addition it is possible to derive distance information from this experiment by measuring the rate of magnetization transfer.

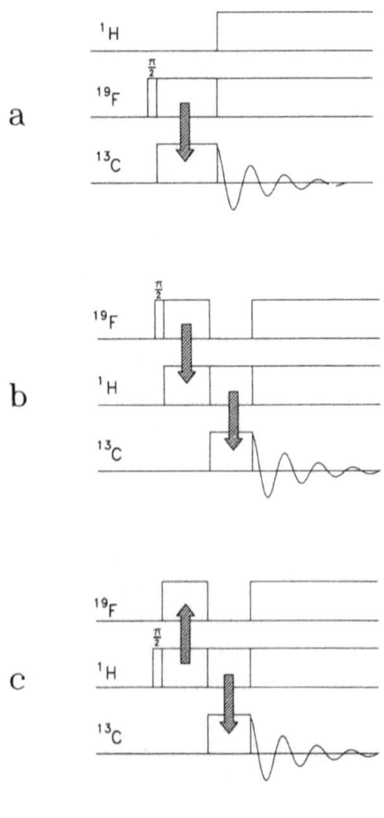


**Fig. 1a–c.** Pulse scheme of the $PMMA/PVF_2$ experiments: **a** $^{19}F-^{13}C$ cross-polarization; **b.** $^{19}F-^{1}H-^{13}C$ double cross-polarization; **c.** $^{1}H-^{19}F$ cross-depolarization followed by $^{1}H-^{13}C$ cross-polarization

Figure 2 shows the $^{13}C$ CPMAS NMR spectra of PMMA and $PVF_2$. They show that the carbonyl and methoxy resonances of PMMA are suitable monitors to study the miscibility of the blends, since they do not overlap with $PVF_2$ carbon-13 resonances. In Fig. 3, the $^{19}F$–$^{13}C$ CPMAS spectra of a physical mixture of PMMA and $PVF_2$ is compared with a PMMA/$PVF_2$ blend. The physical mixture was prepared by mixing the separate components in a mortar after dissolution and precipitation of the homopolymers. PMMA and $PVF_2$ exist in separate domains in this sample, and no PMMA carbon-13 signal was observed when cross-polarizing from fluorines of $PVF_2$. The blend, prepared via coprecipitation, does show PMMA carbon intensities, originating from PMMA carbons in close proximity to $PVF_2$ fluorines.

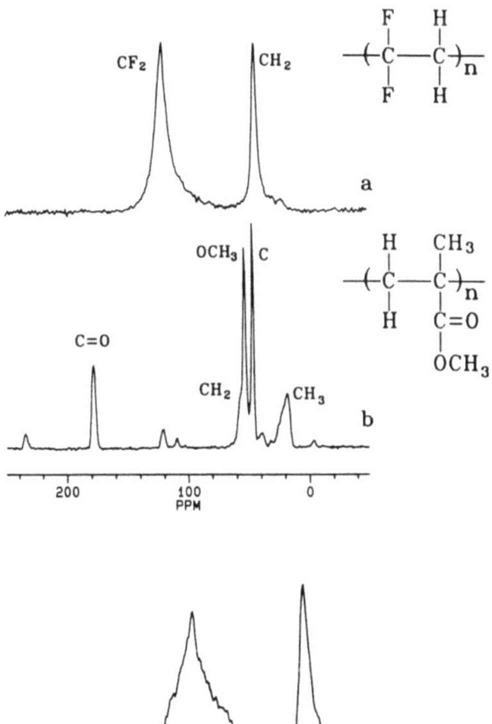


**Fig. 2a, b.** Magic angle spinning $^{13}C$ spectra of (*a*) $PVF_2$, cross-polarized from $^{19}F$ with simultaneous $^1H$ and $^{19}F$ high power decoupling, and (*b*) PMMA, cross-polarized from $^1H$ with $^1H$ high power decoupling

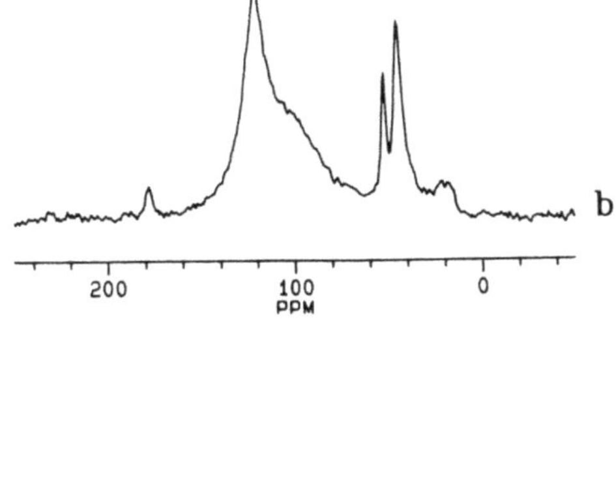


**Fig. 3a, b.** Comparison of $^{19}F$–$^{13}C$ CPMAS spectra of a PMMA/$PVF_2$ 60/40 physical mixture (**a**) and of a blend obtained from coprecipitation (**b**), with a contact time of 5 ms

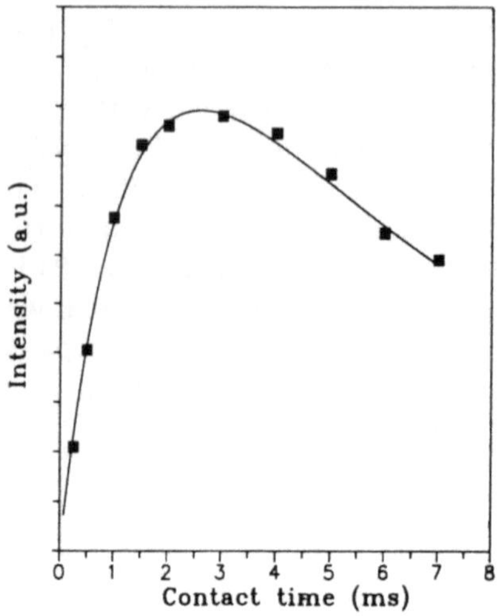


**Fig. 4.** PMMA methoxy carbon intensity as a function of the $^{19}F$–$^{13}C$ cross-polarization time

In order to determine the value of $T_{IS}$, a series of $^{19}F$–$^{13}C$ CP experiments were performed with a variable cross-polarization contact time [53, 54, 55]. In Fig. 4, the intensities of the PMMA methoxy resonance of the PMMA/$PVF_2$ 60/40 blend are plotted as a function of the contact time, together with a calculated curve, obtained by fitting the experimental data with Eq. (3). The $^{19}F$ and $^{13}C$ $T_{1\rho}$'s and the correlation time $\tau_c$ were measured in separate experiments [53, 55]. The average fluorine-carbon distances were found from Eq. (10) and are listed in Table 2 for different PMMA/$PVF_2$ compositions.

In the case of intermolecular polarization transfer in amorphous blends, one would expect the intensity from Eq. (3) as a function of contact time, to be a superposition of curves with different $T_{IS}$ values. By using the relation between the average $I-S$ distance $\langle r \rangle_{IS}$ and $T_{IS}^{-1}$, as expressed by Eq. (10), we can account for a distribution of $T_{IS}$ values by postulating a range of $\langle r \rangle_{IS}$ values $\langle r \rangle_n$. For every $\langle r \rangle_n$ the corresponding $(T_{IS})_n$ is calculated from Eq. (10) and then the CP intensity $I_S(t)$ is fitted to

$$I_s(t) = \sum_n w_n I_{S,n}[t, (T_{IS})_n] \tag{17}$$

where $I_{S,n}$ is given by Eq. (3) with $T_{IS} = (T_{IS})_n$, and $w_n$ are weighting factors. The cross-polarization curves are then calculated for each value of $T_{IS}$ and summed after multiplication with a weighting function $w_n(r)$.

**Table 2.** Average fluorine-carbon distances in the PMMA/$PVF_2$ blends with different weight ratios

| PMMA/$PVF_2$ | 80/20 | 60/40 | 40/60 | 20/80 |
|---|---|---|---|---|
| $^{19}F$–$^{13}CO$ (Å) | $2.7 \pm 0.2$ | $3.1 \pm 0.2$ | $3.0 \pm 0.2$ | $2.6 \pm 0.2$ |
| $^{19}F$–$O^{13}CH_3$ (Å) | $2.7 \pm 0.2$ | $3.0 \pm 0.2$ | $2.9 \pm 0.2$ | $2.7 \pm 0.2$ |

In Fig. 5, the results of attempts to fit the experimental cross-polarization data with Eq. 17 are shown for the $OCH_3$ resonance of the $PMMA/PVF_2$ 60/40 blend. $T_{IS}$ values are calculated for distances ranging from 2.8 Å up to 10 Å. The minimum distance was chosen just below the sum of the van der Waals radii of a fluorine atom in $PVF_2$ and a carbon atom in PMMA. According to the equation for non-bonded interactions given by Loufakis et al. [56] the $^{19}F-^{13}C$ interaction is repulsive only below this minimum distance. Several shapes of the distribution function $w_n(r)$ were used to fit the experimental data, some of which are depicted in Fig. 5a. This figure shows two Gaussian distributions, one with a variance of 0.2 Å (I) and one with a variance of 1.0 Å (II) around an average value of 3.0 Å, together with a spherical distribution (III), which reflects the case where there is no preferential carbon-fluorine distance, and where the probability of the occurrence of a certain distance $r$ increases with $r^2$.

Figure 5b depicts the best fits to the $OCH_3$ carbon intensities of atactic $PMMA/PVF_2$, obtained with the distribution functions of Fig 5a. Good data fits are only obtained by assuming a narrow Gaussian distribution function, around a distance that is close to the average distance found by assuming a single value of $T_{IS}$.

The $^{19}F-^{13}C$ CP experiments on $PMMA/PVF_2$ blends reveal that $PVF_2$ and PMMA are, at least partly, mixed on a molecular scale. The average distance from a fluorine of $PVF_2$ to a carbonyl or methoxy carbon of PMMA is determined to be about 3 Å, independent of the composition of the blend. These

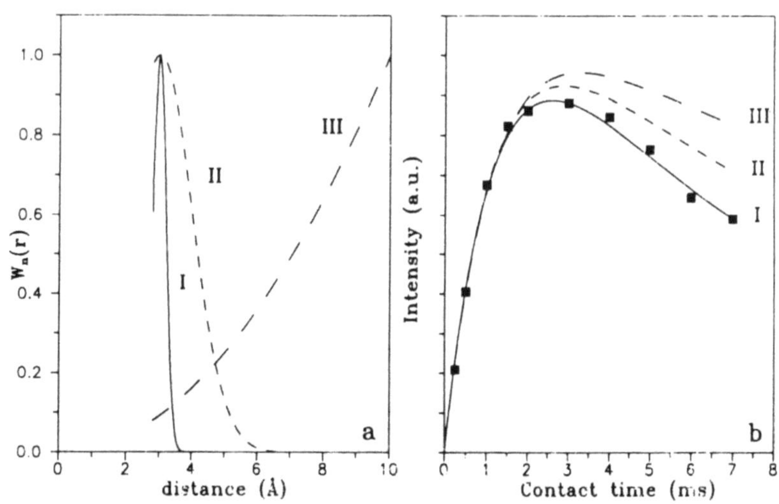


**Fig. 5a.** The weighting functions $w_n(r)$ of Eq. (17); the three curves represent a narrow Gaussian (I), a broad Gaussian (II) and a spherical (III) distribution function of carbons at a distance $r$ from a fluorine aton at $r = 0$; **b** $OCH_3$ intensities of atactic $PMMA/PVF_2$ fitted with the distribution functions of (a)

distances are close to the sum of the van der Waals radii (the reported van der Waals radii are 1.7 Å for a PMMA carbonyl carbon, 1.9 Å for a PMMA ester $CH_3$ group [57] and 1.3 Å for a $PVF_2$ fluorine [58]).

The observed carbon intensities as a function of the cross-polarization time can be fitted either by a single $T_{IS}$ or by a narrow distribution in $T_{IS}$ values. This is indicative of a specific interaction between PMMA and $PVF_2$ segments. This agrees with observations of Léonard et al. [58], who found evidence for strong hydrogen bonding between the carbonyl oxygen of PMMA and the $\alpha$-protons of $PVF_2$ by using FTIR. We do not, however, observe a shift of the carbonyl resonances, as compared to pure PMMA, something one might expect in the case of a strong hydrogen bond. This may be due to the limited spectral resolution, shifts of the order of 0.1–0.2 ppm cannot be detected. Evidence for multiple nearest-neighbor contacts between PMMA and $PVF_2$ was also reported by Saito et al. [59] and Wu [60]. Depolarized light scattering experiments [59] revealed short-range ordering in PMMA/$PVF_2$ due to local chain alignments. The reduced entanglement of PMMA and $PVF_2$ in blends [60] was also attributed to local chain alignments, arising from specific interactions between the polymer segments.

Apart from the distance information obtained from the fluorine-carbon CP experiments, one would like to know the amounts of PMMA and $PVF_2$ that are intimately mixed. In the next sections an attempt will be made to quantify the degree of mixing using different cross polarization techniques.

## 3.3 $^{19}F$–$^{1}H$–$^{13}C$ Double CP

In the $^{19}F$–$^{13}C$ cross-polarization experiments of PMMA/$PVF_2$ blends, PMMA carbon signals are observed from those parts of the PMMA molecules that are close (about 3 Å) to fluorines of $PVF_2$. The intensities of those resonances are small, tempting one to conclude that only a small fraction of the PMMA is close to $PVF_2$. However, due to the short fluorine $T_{1\rho}$ compared to the intermolecular cross-polarization time $T_{IS}$, the build up of PMMA carbon magnetization is restrained. This obscures any conclusions regarding the amount of intimately mixed PMMA and $PVF_2$. The effect of such a short fluorine $T_{1\rho}$ (approximately 1 ms [54]) on the build up of PMMA carbon magnetization as a function of the CP contact time is demonstrated in Fig. 6, where the carbon CP intensities as a function of contact time are compared for two values of the fluorine $T_{1\rho}$.

Such problems can of course be circumvented by performing the experiments at a different temperature. As was shown by Douglass et al. [16] $PVF_2$ exhibits a minimum in fluorine $T_{1\rho}$ at room temperature. Changing the sample temperature is, however, often undesirable in polymer blend systems, since this may result in irreversible changes, such as phase separation and crystallization.

An alternative is to cross-polarize from fluorines to protons. Since the gyromagnetic ratio $\gamma$ of protons is about four times larger than that of carbons,

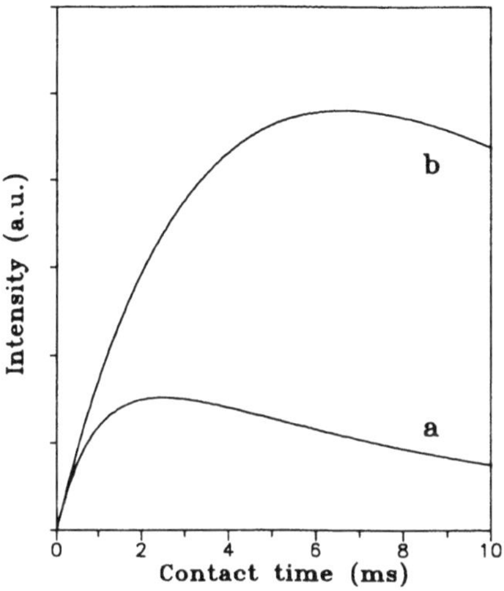


**Fig. 6.** Theoretical cross-polarization curves, calculated with $T_{IS} = 10$ ms, $T^{S}_{1\rho} = 100$ ms and $T^{I}_{1\rho} =$ 1 ms (*a*) and 5 ms (*b*)

the fluorine-proton magnetization transfer is therefore expected to be at least sixteen times faster than from fluorines to carbons, resulting in a more favorable $T_{IS}/T^{I}_{1\rho}$ ratio and a more efficient build up of proton magnetization, as compared to carbon magnetization. Of course; we still want to discriminate between PMMA and $PVF_2$ proton magnetization and since the proton resonances of PMMA and $PVF_2$ overlap, the proton magnetization is subsequently transferred to carbons via proton-carbon cross-polarization (see Fig. 1b). This last step is much more efficient than $^{19}F-^{13}C$ transfer because of the longer proton $T_{1\rho}$ (4.5–7 ms, depending on the composition of the blend).

A comparison between carbon-13 spectra, obtained via $^{19}F-^{13}C$ cross-polarization, and via $^{19}F-^{1}H-^{13}C$ double cross polarization, is shown in Fig. 7. As can be seen from these spectra the double transfer experiment indeed results in much larger PMMA carbon signals. The double CP spectrum is obtained by averaging 1000 scans, while the lower spectrum resulted from 5000 scans. The spectra have been arbitrarily scaled for representation purposes. Both experiments, however, detect the same amount of $PVF_2$, so the intensities of the $CF_2$ resonances in both spectra should be approximately equal. The faster fluorine-proton transfer, visualized in Fig. 8, plus additional proton spin diffusion from $PVF_2$ to PMMA and finally $^{1}H-^{13}C$ cross-polarization of PMMA, increases the PMMA signal.

An interesting detail in the $^{19}F-^{1}H-^{13}C$ double CP spectra is the appearance of the quaternary $^{13}C$ resonance of PMMA. This resonance is not detected separately in $^{19}F-^{13}C$ CP spectra, due to overlap of this resonance with the $CH_2$ resonance of $PVF_2$, and because it is so weak. Since the quaternary carbon is located in the backbone of the PMMA chain, its distance to fluorines is expected to be larger than the distances found for carbonyl and methoxy carbons.

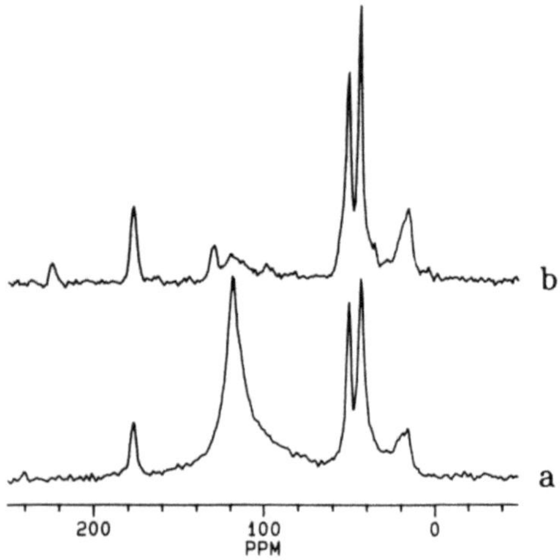


**Fig. 7.** (*a*) $^{19}F-^{13}C$ CPMAS spectrum of PMMA/$PVF_2$ 60/40, obtained with a contact time of 2 ms; (*b*) $^{19}F-^{1}H-^{13}C$ double CP MAS spectrum of the same blend, using a $^{19}F-^{1}H$ contact time of 0.4 ms and a $^{1}H-^{13}C$ contact time of 2 ms

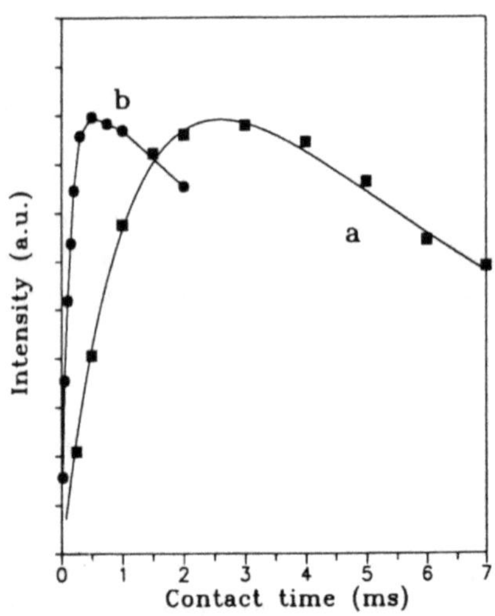


**Fig. 8.** PMMA methoxy carbon intensities as a function of the $^{19}F-^{13}C$ cross-polarization time (*a*), and of the $^{19}F-^{1}H$ cross-polarization time (*b*). The intensities are scaled arbitrarily

Carbon-13 spectra obtained with the double CP experiment with increasing $^{19}F-^{1}H$ CP contact time, show an efficient build up of PMMA carbon magnetization. Unfortunately, the double cross polarization experiments can give us only qualitative information about $PVF_2$-PMMA distances. A quantitative analysis is complicated because of the effect of proton spin diffusion. As soon as proton magnetization is generated by $^{19}F-^{1}H$ CP in a domain where PMMA and $PVF_2$ are intimately mixed, the proton magnetization will diffuse out of this region to PMMA protons that are further away from $PVF_2$ fluorines than the 3 Å estimated from the $^{19}F-^{13}C$ CP experiment. The observed PMMA magnetization stems therefore from PMMA molecules that are, on average, at larger distances from $PVF_2$ than in the case of the direct $^{19}F-^{13}C$ CP experiment.

The $^{19}F-^{1}H-^{13}C$ CP technique is able to provide qualitative proof of molecular miscibility up to approximately 15–20 Å and will be useful in materials where fluorine-carbon CP fails to indicate spatial proximity because a too large average fluorine-carbon distance results in a very slow polarization transfer. A modification of the $^{19}F-^{1}H-^{13}C$ experiment described above can however be used to give quantitative results about the fraction of PMMA units that is within proton spin diffusion distances from $PVF_2$ units. This proton-fluorine cross depolarization experiment will be described in the next section.

## 3.4 $^{1}H-^{19}F$ Cross-Depolarization

### 3.4.1 Cross-Depolarization

Instead of using the fluorines as a source of magnetization, as was the case in the above mentioned techniques, we will now use the fluorine reservoir as a sink, through which proton magnetization can disappear.

If, after creation of spin-locked proton magnetization in the whole sample, a fluorine RF field is turned on, its strength adjusted to the proton-fluorine Hartmann-Hahn condition, then all the protons that are dipolar coupled to fluorines will lose their magnetization to the lattice, via the fluorines. In a second step the remaining proton magnetization is transferred to carbons via $^{1}H-^{13}C$ CP, and the carbon magnetization is detected (see Fig. 1c). The observed PMMA carbon magnetization is then from parts of PMMA molecules, whose protons do *not* have a dipolar interaction with fluorine. In other words, only those parts of PMMA molecules that are remote from $PVF_2$ units are observed.

In performing this $^{1}H-^{19}F$ cross depolarization experiment, the short fluorine $T_{1\rho}$ is an advantage in that it makes the fluorine reservoir act as a good sink for proton magnetization. In order to make sure that this sink functions properly for longer depolarization contact times, one has to prevent fluorine magnetization from building up. This can be done, either by applying a phase modulated RF field or by frequently switching the phase of the RF field by 180°, as was done here. In order to determine the amount of PMMA that is close to $PVF_2$, a blank experiment was performed, in which no fluorine RF field was used. If we denote the signal from the depolarization experiment $S_A$, and the signal from the blank experiment $S_B$, we can write the observed signals as:

$$S_A(t) = S_0 f_{mix} \exp\left(-\frac{t}{T_{1\rho}}\right) D(t) + S_0(1 - f_{mix}) \exp\left(-\frac{t}{T_{1\rho}}\right) \tag{18}$$

$$S_B(t) = \{S_0 f_{mix} + S_0(1 - f_{mix})\} \exp\left(-\frac{t}{T_{1\rho}}\right) \tag{19}$$

where $S_0$ is the total proton magnetization, $f_{mix}$ is the fraction of protons close

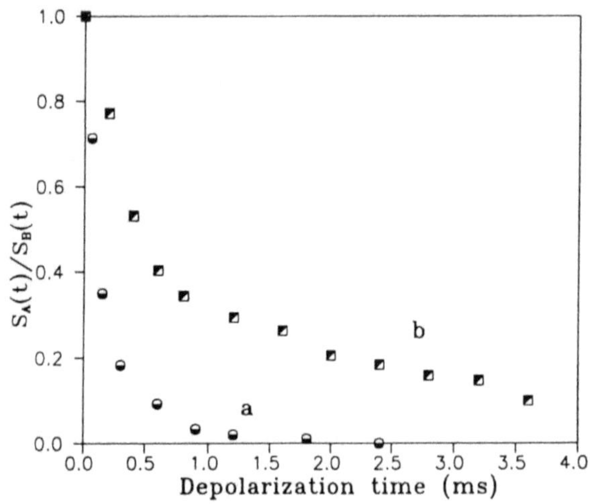


**Fig. 9.** Carbon intensities of the $CF_2$ resonance as a function of the $^1H-^{19}F$ contact time (depolarization time), (*a*) $PVF_2$; (*b*) PMMA/$PVF_2$ 60/40

to fluorines and $D(t)$ is a decaying function which describes the loss of magnetization due to depolarization to the fluorine sink. By dividing both signals, we have corrected for loss of magnetization to the lattice, characterized by the poton $T_{1\rho}$:

$$\frac{S_A(t)}{S_B(t)} = f_{mix} D(t) + (1 - f_{mix}) \tag{20}$$

By transferring at time $t$ the proton magnetization to the carbon nuclei, the proton signal $S_A(t)/S_B(t)$ can be mapped out via the carbon resonances of PMMA or $PVF_2$.

In Fig. 9a the data for pure $PVF_2$, detected via $CF_2$, are represented in this way. The value of $S_A/S_B$ drops to 0, which is expected since all the protons in $PVF_2$ are close to fluorines, and therefore $f_{mix} = 1$. The effect of proton spin diffusion is clearly shown in Fig. 9b, where the data obtained for the $CF_2$ resonances in the PMMA/$PVF_2$ 60/40 blend are depicted. In pure $PVF_2$, all the proton magnetization has disappeared after 1.2 ms, while for the same protons in the blend this process is much slower, as can be seen by comparing Figs. 9a and b. As soon as some proton magnetization is destroyed in a regiòn close to the fluorine nuclei, magnetization will be transported from PMMA protons in outer regions to this area, due to spin diffusion driven by the magnetization concentration gradient. In spite of the effect of this spin diffusion, the proton magnetization of $PVF_2$ units in a blend, detected via $CF_2$ carbons, decays to zero in about 3–5 ms, depending on the composition of the blend. This is in contrast to the results for the PMMA proton magnetization (see Fig. 10), detected via the C=O or $OCH_3$ resonance, which levels off to a nearly constant value at 3.5 ms. This apparent constant level is attributed to the proton magnetization of the fraction 1-$f_{mix}$ of PMMA that is not closely mixed with $PVF_2$. In the absence of proton spin diffusion PMMA units close to and PMMA

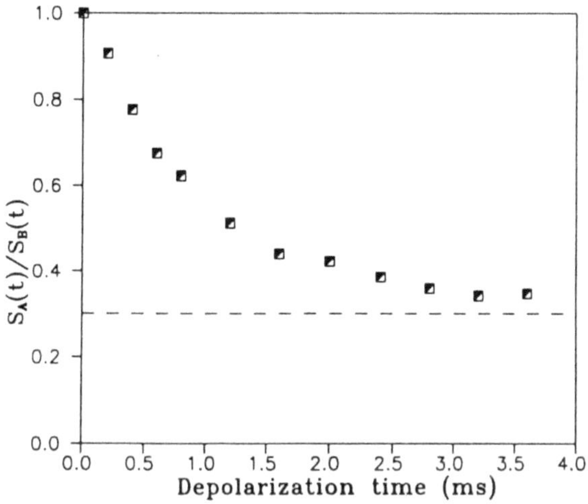


**Fig. 10.** Carbon intensities of the C=O resonance of PMMA in the PMMA/$PVF_2$ 60/40 blend as a function of the $^1H-^{19}F$ contact or depolarization time

units remote from $PVF_2$ can be distinguished by the distance aspect of $^1H-^{19}F$ cross polarization. Proton spin diffusion complicates this distinction but clearly, the ability of proton spin diffusion to transfer magnetization from PMMA protons to $PVF_2$ protons can also be used to distinguish well-mixed PMMA from non-mixed PMMA.

### *3.4.2 The Diffusion Model*

The transfer of proton magnetization in solids via the flip-flop term *B* of the dipolar spin Hamiltonian of Eq. (3) may be described by a diffusion equation:

$$\frac{\partial m(r,t)}{\partial t} = D\frac{\partial^2 m(r,t)}{\partial r^2} + f(r)m(r,t) \tag{21}$$

This equation is very similar to Eq. (12) except that a spatially dependent function $f(r)$ has been added which describes the depolarization.

The blend is assumed to consist of four different phases:

1. An isolated $PVF_2$ phase I, i.e. on the timescale of the experiment no PMMA proton magnetization can travel into this domain through spin diffusion.
2. A PMMA/$PVF_2$ phase II in which PVF and PMMA are intimately mixed, and where the main transport of magnetization from protons to fluorines is provided by direct $^1H-^{19}F$ depolarization.
3. A PMMA phase III which is close to a mixed phase II, so that its local magnetization density is influenced, through spin diffusion, by the loss of proton magnetization in the mixed phase.
4. An isolated PMMA phase IV whose magnetization is not affected by depolarization, nor by spin diffusion.

For the four phases mentioned above, the local magnetization density will be determined by several parameters but, because the experimental data are corrected for direct loss of proton magnetization to the lattice via $T_{1\rho}$ processes, we do not have to include the effect of the proton $T_{1\rho}$.

The isolated $PVF_2$ phase I will lose its proton magnetization to fluorines through cross depolarization, with a rate constant $T_{HF}^{-1}$. This rate constant will be approximately equal to that observed in pure $PVF_2$. The magnetization of the isolated PMMA, phase IV, will be assumed constant, since on the timescale of the experiment no magnetization will diffuse out of this region.

The proton magnetization, present in the mixed phase II, will decay due to cross depolarization to fluorines, with a rate constant $(T_{HF}^{eff})^{-1}$. In addition there will be a supply of proton magnetization, through spin diffusion from the neighboring PMMA phase III.

On the basis of the experimental data for the $CF_2$ resonance it was found that $(T_{HF}^{eff})^{-1}$ does not equal $T_{HF}^{-1}$, the rate constant for cross depolarization in pure $PVF_2$. One obvious reason for this is the presence of PMMA protons, close to fluorines, representing a larger total proton reservoir, as compared to pure $PVF_2$, leading to a slower destruction of proton polarization.

The proton magnetization, present in the different phases, can be monitored separately, since the proton magnetization is subsequently transferred to carbon magnetization via $^1H-^{13}C$ cross polarization. In the observed carbon-13 spectra, the $CF_2$ resonance will therefore reflect $PVF_2$ molecules present in phases I and II. The C=O, $OCH_3$, and $CH_3$ resonances of PMMA, will reflect the PMMA molecules present in phases II, III and IV. The fraction of isolated PMMA (phase IV), however, will appear in the plots of magnetization vs. intensity, as an offset from zero. From the observed value of the magnetization for long depolarization times, this fraction can be found directly from the plot, as can be seen in Fig. 10.

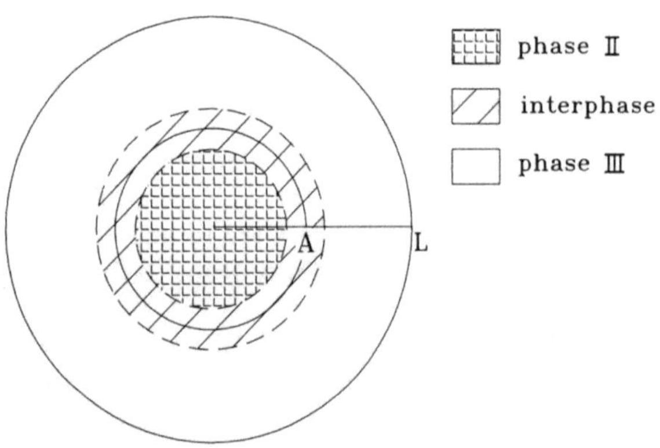


**Fig. 11.** Distribution of phases II and III as used in the spin diffusion model

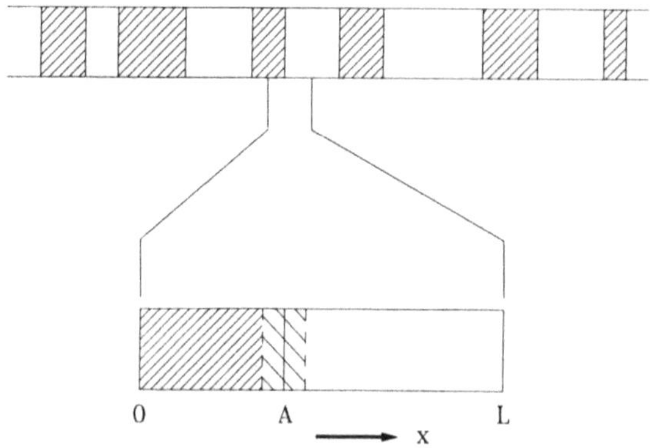


**Fig. 12.** One-dimensional representation of the distribution of domains containing phase II (*shaded area*) or phase III (*non-shaded areas*)

For the mixed phase II and the neighboring PMMA phase III the dependence of the magnetization on the depolarization time can be described by the spin diffusion equation. As a simple approximation we assume radial spin diffusion, from an outer sphere containing PMMA, into an inner sphere consisting of phase II, (Fig. 11).

In Fig. 12, a one-dimensional distribution of domains, consisting of phase II and III, is shown. The shaded areas represent mixed domains (phase II); the non-shaded domains contain PMMA (phase III). The radii of the domains are random, and independent of each other. The actual calculations are performed in three dimensions.

In order to find the boundary conditions for the spin diffusion equation, we choose a 'repeat unit', as depicted in the expansion of Fig. 12. In the mixed domain ($-A \leqq x \leqq A$), there will always be a point through which no flow of magnetization takes place, since magnetization flow from the right-hand and left-hand side of this point will cancel each other. If, for simplicity, we choose this point at $x = 0$, then our first boundary condition is:

$$\frac{\partial m(x,t)}{\partial x} = 0 \quad \text{for } x = 0 \tag{22}$$

For the same reasons a point $L$ in the domain of phase III can be defined, through which there is no magnetization flow, since flow will go either in the direction of a sink to the left of this point, or to a sink to the right of this point. The second boundary condition is therefore:

$$\frac{\partial m(x,t)}{\partial x} = 0 \quad \text{for } x = L \tag{23}$$

Via the depolarization function $f(r) = f(x)$ in the diffusion equation (Eq. 21) one takes into account that the depolarization takes place only in the mixed domain:

$$f(x) = \begin{cases} -k & 0 \leqq x \leqq A \\ 0 & A \leqq x \leqq L \end{cases} \tag{24}$$

where $k$ is the depolarization rate constant $(T_{HF}^{eff})^{-1}$. A more realistic approach, however, is to introduce a continuous transition between the mixed phase and the neighboring PMMA domain. This interface, depicted by the differently shaded areas in Figs. 11 and 12, is characterized by a decaying value of $|k|$ in the direction of the PMMA domain. For mathematical convenience only, a hyperbolic tangent was chosen, which mimicks an interfacial width of about 3 Å. This will closely resemble the $x^{-6}$ dependence of the proton-fluorine CP rate [54]. We therefore replace $f(x)$ in Eq. (24) by the following expression:

$$f(x) = -\tfrac{1}{2}k[1 - \tanh\{b(x - A)\}] \tag{25}$$

in which $b$ can be varied, in order to adjust the width of the interface to about 3 Å. The initial condition for the differential equation arises from the fact that at $t = 0$ proton magnetization is present everywhere in the sample, hence:

$$C = C_0 \quad \text{for all } x \text{ at } t = 0 \tag{26}$$

Using the boundary and initial conditions given above, the spin diffusion equation can be evaluated numerically [61].

One also has to take into account the diffusion that occurs during the last step of the experiment, i.e. the transfer of proton polarization to carbon nuclei. Under the assumption that cross polarization from protons to carbons is equally fast for all carbons, the spin diffusion during $^{1}H-^{13}C$ cross polarization can be described by:

$$\frac{\partial m(x,t)}{\partial t} = D\frac{\partial^2 m(x,t)}{\partial x^2} \tag{27}$$

which is evaluated using the same boundary conditions as in the case of cross depolarization. The initial distribution is given by the concentration at site $x$, evaluated at depolarization time $t$.

The carbon intensities as a function of depolarization time $t$ are obtained by integrating over $x$. For the $CF_2$ resonance the intensities are given by integrating over the sink, i.e. by integrating over $x$ from 0 to $(A - 1.5)$ Å. The observed PMMA intensities result if one integrates over $x$ from 0 to $L$. In the actual calculations, the mixed phase was contained in a sphere with radius $A$, surrounded by an outer shell from $r = A$ to $r = L$, containing PMMA (see Fig. 11). Integration was carried out over the inner sphere in order to obtain $CF_2$ intensities, and from $r = 0$ to $r = L$ for PMMA intensities.

### 3.4.3 Evaluation of the Diffusion Model

The model above described can be used to explain the experimental depolarization data by numerically evaluating the spin diffusion equations. The spin diffusion coefficient can be expressed in terms of the local dipolar field of the protons [54, 62, 63, 32],

$$D = \frac{1}{2}\sqrt{\frac{\pi}{60}}\langle a^2 \rangle T_2^{-1} \tag{28}$$

where $T_2$ is the proton transverse spin-lattice relaxation time, and $\langle a^2 \rangle$ is the average of the square of the distances between adjacent protons in PMMA. In this evaluation we assumed that spin diffusion proceeds mainly between PMMA protons and not in the $PVF_2$ segments. The use of a single diffusion coefficient is therefore justified.

The proton $T_2$'s are obtained from Hahn echo experiments [54] and $\langle a^2 \rangle$ is estimated from the average over the squared distances between adjacent protons in PMMA, obtained from a molecular modeling study on a strand of isotactic PMMA [54, 64]. The value of $T_{HF}$ is obtained from pure $PVF_2$ and $(T_{HF}^{eff})^{-1}$ is determined from the depolarization data of the $CF_2$-resonance in the blend at short depolarization times. The calculated depolarization curves are then fitted to the experimental data by varying $A$ and $L$. In order to fit the curves to the experimental data obtained from PMMA carbons, one has to take into account the amount of PMMA that is not mixed (phase IV). This fraction was derived from the asymptotic behavior of the depolarization data at longer cross depolarization times. In Fig. 13, the experimental data for the carbonyl resonance of PMMA/$PVF_2$ 60/40 is shown, together with the calculated curve for $A = 6$ Å and $L = 12$ Å, and a fraction of 30% PMMA contained in phase IV. The calculated curves are scaled by dividing the integrated magnetization by its value at time $t = 0$: $M/M_0$. Table 3 lists the results for PMMA/$PVF_2$ blends with different weight ratios. The depolarization data for $PVF_2$, as obtained from the $CF_2$ carbon resonance, are composed of a spin diffusion curve, which accounts for $PVF_2$ contained in phase II, and an exponential decay curve, with

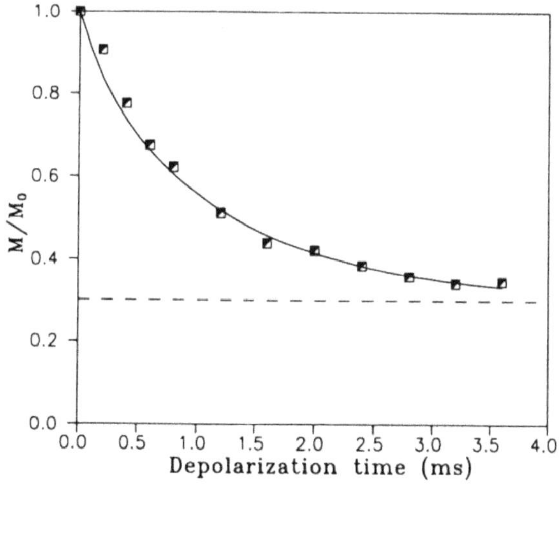


**Fig. 13.** Experimental and fitted data for the C=O resonance of the PMMA/$PVF_2$ 60/40 blend, fitted with $A = 6$ Å and $L = 12$ Å

**Table 3.** Isolated fractions of PMMA and $PVF_2$, as determined from the cross-depolarization data

| $PMMA/PVF_2$ | 20/80 | 40/60 | 60/40 | 80/20 |
|---|---|---|---|---|
| % PMMA non-mixed | 0 | 10 | 30 | 30 |
| % $PVF_2$ non-mixed | 75 | 35 | 15 | 10 |

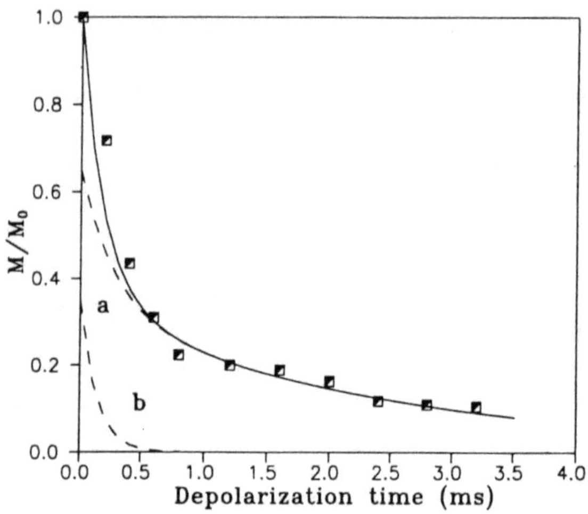


**Fig. 14.** Experimental data from the $CF_2$ resonance of $PMMA/PVF_2$ 40/60 blend. The solid line is the sum of a calculated depolarization curve with $A = 6$ Å and $L = 15$ Å (*a*) to account for mixed $PVF_2$ and an exponential decay (*b*) representing the isolated $PVF_2$

time constant $T_{HF}^{-1}$, representing $PVF_2$ that is not mixed (phase I). This is clearly demonstrated in Fig. 14 for the $PMMA/PVF_2$ 40/60 blend, in which 35% of $PVF_2$ is contained in phase I. In this figure the curves for phase II and I are shown separately, together with their sum. The dashed line (a) is a calculated depolarization curve with $A = 6$ Å and $L = 15$ Å, multiplied by 0.65, to account for 65% of $PVF_2$ in the mixed phase II. Curve *b* is the depolarization curve of 35% of $PVF_2$ that is not mixed and therefore not influenced by spin diffusion. The solid line is the sum of the curves *a* and *b*. Both the fractions of isolated PMMA and $PVF_2$ are determined in a way that is independent of the choice of a spin diffusion model.

The data obtained for the different PMMA resonances are, within experimental error, all equal, as expected. The results obtained for blends with different $PMMA/PVF_2$ ratios are shown in Table 3. It appears that the data for PMMA resonances from different compositions of the blend are equally well described by the curves with $A = 6$–$7$ Å and $L = 12$–$13$ Å. This indicates that the nature of the miscible domains is the same for all blends in the composition range studied. Good data fits were obtained for the $CF_2$ carbon data of $PVF_2$, with $A = 6$–$7$ Å and $L = 12$–$17$ Å. Except for the $PMMA/PVF_2$ 20/80 blend, the values for *L*, found from $CF_2$, are slightly larger than the *L* values found from the corresponding PMMA resonances.

### 3.4.4 Discussion

The cross depolarization experiments provide us with information of two kinds. First, we are able to determine directly the amounts of PMMA and $PVF_2$ that are not intimately mixed. These fractions are independent of any spin diffusion model. The criterion used here for distinguishing mixed and non-mixed $PVF_2$ and PMMA, is based on spatial separation by distances that cannot be bridged by proton spin diffusion within the time of the depolarization. Secondly, the introduction of a spherical spin diffusion model leads to the estimation of the parameters $A$ and $L$. Although the proposed model is clearly too simple to describe the complexity of a polymer blend, it is surprising how well the calculated curves describe the experimental data and how sensitive [54] they are to changes in $A$ and $L$. However, these parameters arise solely from a mathematical model and it is not clear how they are related to geometrical dimensions in the blend. It seems likely that the value of the $A$ parameter found from the fitting procedure simply expresses the fact that $^1H-^{19}F$ cross polarization over distances larger than $A$ Å is too slow to be noticed. That is after all the way $A$ is defined. $L$ is defined as the distance over which proton spin diffusion is in operation during depolarization. The values found for $L$, however, are much smaller than the proton spin diffusion length. Rather than speculating about the physical meaning of $A$ and $L$, we propose to use the results from this model as an analytical tool when comparing blends with different composition ratios, different constituents or when studying the response of a blend to external influences such as aging or heating.

The techniques discussed in this section provide detailed information on miscibility at a molecular scale. They have also been applied to study conformational effects and dynamic processes in PMMA/$PVF_2$ blends. Eijkelenboom et al. [55] studied the effect of PMMA microstructure on miscibility with $PVF_2$. These results show that the nature of the mixing of isotactic, syndiotatic and atactic PMMA with $PVF_2$ is similar, as revealed through equal $^{19}F-^{13}C$ distances from the $^{19}F-^{13}C$ cross polarization experiment. The total amounts of intimately mixed PMMA and $PVF_2$ differed, however, and was found to increase in the order syndiotactic, atactic, isotactic PMMA.

The crystallization behavior of $PVF_2$ in the PMMA/$PVF_2$ blend was investigated using the $^1H-^{19}F$ cross depolarization technique [65]. The amount of mixed material was found to decrease as a function of the annealing time at elevated temperatures, indicating a phase separation in the blend. By comparing these data to those obtained from DSC measurements, the phase separation phenomena could be ascribed to crystallization of $PVF_2$ in the blend. In addition evidence was found for an amorphous interphase, surrounding the $PVF_2$ crystallites, in accordance with suggestions of the existence of such an interphase by Hahn et al. [48].

In another study by Maas et al. [66] the interdiffusion of PMMA and $PVF_2$ was studied by the $^1H-^{19}F$ cross depolarization technique. The observed decrease in the isolated PMMA and $PVF_2$ fractions in alternatingly stacked sheets of $PVF_2$ and PMMA as a function of the time during which the polymers are melted at 190 °C, allowed the determination of the intrinsic diffusion coefficients of both components.

## 4 Conclusion

The foregoing sections have shown that (solid state) NMR is an unique technique for studying microscopic miscibility in polymer blends. No other technique at present can give such detailed information about the microscopical structure. Many of the techniques discussed, however, are not yet suitable for routine measurements. The authors believe that this will change in the foreseeable future.

## 5 References


1. Olabisi O, Robeson LM, Shaw MT (1979) Polymer-polymer miscibility. Springer Series in Solid-state Sciences. Springer, Berlin Heidelberg New York
2. Pines A, Gibby MC, Waugh JS (1973) Chem Phys 59: 569
3. Andrew ER (1971) In: Emsley JW, Feeney F, Sutcliffe LH (eds) Progress in NMR spectroscopy, Pergamon, Oxford, vol 8, p 1
4. Haeberlen U (1979) In: Waugh JS (ed), Advances in Magnetic Resonance, Suppl. 1
5. Gerstein BC, Dybowski C (1985) Transient techniques in NMR of solids. Academic, New York
6. Noggle JH, Schirmer RE (1971) The nuclear Overhauser effect, Academic, New York
7. Wind RA, Duijvestijn, van der Lugt C, Manenschijn A, Vriend J (1985) J Progr Nucl Magn Reson Spectroscopy 17: 33
8. Raleigh DP, Levitt MH, Grifffn RG (1988) Chem Phys Letters 146: 71
9. Belfiore LA, Graham H, Veda E, Wang V (1992) Polymer Internat 28: 81
10. Zumbulyadis N, Landry CJT, Long TE (1993) Macromolecules 26: 2647
11. McGrath KJ, Roland CM (1992) Macromolecules 25: 166
12. Graessley WW, Krishnamoorti R, Balsara NP, Fetters LJ, Lohse DJ, Schulz N, Sissano JA (1993) Macromolecules 26: 1137
13. Mehring M (1983) In: Diehl P, Fluck E, Kosfeld R (eds) NMR, basic principles and progress. Springer Berlin Heidelberg, New York, vol 12, 2nd edn 1983
14. Demco DE, Tegenfeldt J, Waugh JS (1975) Phys Rev B, 11: 4133
15. Slichter CP (1990) Principles of Magnetic Resonance. Springer, Berlin Heidelberg New York
16. Douglass DC, McBrierty VJ (1978) Macromolecules 11: 766
17. Kögler G, Mirau PA (1992) Macromolecules 25: 598
18. Mirau PA, Heffner SA, Kögler G, Bovey FA (1991) Polymer Internat. 26: 29
19. Kelts LW, Landry CJT, Teegarden DM (1993) Macromolecules 26: 2941
20. White JL, Mirau PA (1993) Macromolecules 26: 3049
21. White JL, Mirau PA (1993) Macromolecules (in press)
22. Linder M, Henrichs PM, Hewitt JM, Massa DJ (1985) J Chem Phys 82: 1585
23. Tang P, Reimer JA, Denn MM (1993) Macromolecules 26: 4269
24. Crank J (1956) Mathematics of diffusion, Clarendon Oxford
25. Clauss J, Schmidt-Rohr K, Spiess HW (1993) Acta Polym. 44: 1
26. Ernst RR, Bodenhausen G, Wokaun A (1987) Principles of nuclear magnetic resonance in one and two dimensions, Clarendon, Oxford
27. Caravatti P, Neuenschwandler P, Ernst RR, (1986) Macromolecules 19: 1895

28. Jeener J, Meier BH, Bachmann P, Ernst RR (1979) J Chem Phys 71: 462
29. Caravatti P, Neuenschwandler P, Ernst RR (1986) Macromolecules 19: 1889
30. Henrichs PM, Tribone J, Massa DJ, Hewitt JM (1988) Macromolecules 21: 1282
31. Goldman M, Shen L (1966) Phys Rev 144: 321
32. Havens JR, VanderHart DL (1985) Macromolecules 18: 1663
33. Stejskal EO, Schaefer J, Sefcik MD, McKay RA (1981) Macromolecules 14: 275
34. Maresch GG, Kendrick RD, Yannoni CS, Galvin ME (1988) Macromolecules 21: 3525
35. Afeworki M, McKay RA, Schaefer J (1992) Macromolecules 25: 4084
36. VanderHart DL, Manders WF, Stein RS, Herman W (1987) Macromolecules 20: 1726
37. Walton JH, Miller JB, Roland CM (1992) J. Polymer Sci: Part B: Polymer Physics 30: 527
38. Walton JH, Miller JB, Roland CM, Nagode JB (1993) Macromolecules 26: 4052
39. Mansfeld M, Veeman WS (1993) Chem Phys Letters 213: 153
40. Sarkar SN, Komoroski RA (1992) Macromolecules 25: 1420
41. Cory DG, de Boer JC, Veeman WS (1989) Macromolecules 22: 1618
42. Noland JS, Hsu NN-C, Saxon R, Schmitt JM (1971) Adv Chem Ser, 99: 15
43. Nishi T, Wang TT (1975) Macromolecules 8: 909
44. Bernstein RR, Gruz CA, Paul DR, Barlow JW (1977) Macromolecules 10: 681
45. Wendorff JH (1980) J Polym Sci, Polym Lett Ed, 18: 439
46. Roerdink E, Challa G (1978) Polymer 19: 173
47. Léonard C, Halary JL, Monnerie L (1988) Macromolecules 12: 1265
48. Hahn BR, Hermann-Schonherr O, Wendorff JH (1987) Polymer 28: 201
49. Ward TC, Lin TS (1983) Polym. Prep. 24: 136
50. Ward TC, Lin TS (1984) In: Hahn, CD (ed) Polymer blends and composites in multiphase systems, Advances in Chemistry. American Chemical Society, Washington, DC, vol 206 p 59
51. Tékély P, Lauprêtre F, Monnerie L (1985) Polymer 26: 1081
52. Grinsted RA, Koenig JL (1988) Polymer Prepr 29: 15
53. Klein Douwel CH, Maas WEJR, Veeman WS, Werumeus Buning GH, Vankan JMJ (1990) Macromolecules 23: 406
54. Maas WEJR, van der Heijden WAC, Veeman WS, Vakan JMJ, Werumeus Buning GH (1991) J Chem Phys 95: 4698
55. Eijkelenboom APA, Maas WEJR, Veeman WS, Werumeus Buning GH, Vankan JMJ (1992) Macromolecules 25: 28
56. Loufakis K, Miller KJ, Wunderlich B (1986) Macromolecules 19: 1271
57. Bosscher F, ten Brinke G, Eshuis A, Challa G (1982) Macromolecules 15: 1364
58. Léonard C, Halary JL, Monnerie L (1985) Polymer 26: 1507
59. Saito H, Matsuura M, Okada T, Inoue T (1989) Polymer J. 21: 357
60. Wu S (1987) J Polym Sci, Part B: Polym Phys 25: 557
61. Ames WF (1977) Numerical methods for partial differential equations, 2nd edn. Academic, New York
62. Cheung TTP, Gerstein BC (1981) J Appl Phys 52: 5517
63. Cheung TTP (1981) Phys Rev B, 23: 1404
64. Kusanagi H, Tadokoro H, Chatani Y (1976) Macromolecules 9: 531
65. Papavoine CHM, Maas WEJR, Veeman WS, Werumeus Buning GH, Vankan JMJ (1993) Macromolecules 26: 6611
66. Maas WEJR, Papavoine CHM, Veeman WS, Werumeus Buning GH, Vankan JMJ, in press
67. Takegoshi K, Hikichi K (1991) J Chem Phys 94: 3200
68. Le Menestrel C, Kenwright AM, Sergot P, Lauprêtre F, Monnerie L (1992) Macromolecules 25: 3020
69. Gobbi GC, Sivestri R, Russell TP, Lyerla JR, Fleming WW (1987) Polym Sci, Polymer Lett ED 25: 61
70. Parmer JF, Dickinson LC, Chien JCW, Porter RS (1987) Macromolecules 20: 2308
71. Chu CW, Dickinson LC, Chien JCW (1988) Polymer Bull. 19: 265
72. Zhang X, Natansohn A, Eisenberg A (1990) Macromolecules 23: 412
73. Zhang C, Wang P, Jones AA, Inglefield PT (1991) Macromolecules 24: 338
74. Chin YH, Zhang C, Wang P, Inglefield PT, Jones AA, Kambour RP, Bendler JT, White DM (1992) Macromolecules 25: 3031
75. Campbell GC, VanderHart DL, Feng Y, Han CC (1992) Macromolecules 25: 2107
76. Stejskal EO, Schaefer J, Sefcik MD, McKay RA (1981) Macromolecules 14: 188
77. Zhang X, Takegoshi K, Hikichi K (1991) Macromolecules 24: 5757
78. Schenk W, Reichert D, Schneider H (1990) Polymer 31: 329

79. Brosseau C, Guillermo A, Cohen-Addad JP (1992) Polymer 33: 2076
80. Zawada JA, Ylitalo CM, Fuller GG, Colby RH, Long TE (1992) Macromolecules 25: 2896
81. Marco C, Fatou JG, Gomez MA, Tanaka H, Tonelli AE (1990) Macromolecules 23: 2183
82. Parmer JF, Dickinson LC, Chien JCW (1989) Macromolecules 22: 1078
83. Spevacek J, Schneider B, Straka J (1990) Macromolecules 23: 3042
84. Albert B, Jérôme R, Teyssié P, Smyth G, Boyle NG, McBrierty VJ (1985) Macromolecules 18: 388
85. Qin C, Pires ATN, Belfiore LA (1991) Macromolecules 24: 666
86. Zhang X, Takegoshi K, Hikichi K (1992) Macromolecules 25: 2340
87. Kwei TK, Dai YK, Lu X, Weiss RA (1993) Macromolecules 26: 6563
88. Immirzi B, Malinconico M, Martuscelli E (1992)
89. Immirzi B, Malinconico M, Martuscelli E, Paci M, Segre A (1992) Makromolekulare Chemie 13: 231
90. Schantz S, Ljungqvist N (1993) Macromolecules 26: 6517
91. Iriarte M, Espi E, Etxeberria Valero M, Fernández-Berridi MJ, Iruin JJ (1991) Macromolecules 24: 5546
92. Natansohn A, Simmons A (1989) Macromolecules 22: 4426
93. Miller JB, McGrath KJ, Roland CM, Trask CA, Garroway AN (1990) Macromolecules 23: 4543
94. Masson J-F, Manley RStJ (1992) Macromolecules 25: 589
95. Simmons A, Natansohn A (1992) Macromolecules 25: 3881
96. Natansohn A, Lacasse M, Banu D, and Feldman D (1990) J Appl Polymer Sci 40: 899
97. Zhang X, Takegoshi K, Hikichi K (1992) Polymer 33: 712
98. Huo PP, Cebe P (1993) Macromolecules 26: 5561
99. Schmidt-Rohr K, Clauss J, Blümich B, Spiess W (1990) Magnetic Resonance in Chemistry 28: S3
100. Dickinson LC, Yang H, Chu CW, Stein RS, Chien JCW (1987) Macromolecules 20: 1757
101. Grobelny J, Rice DM, Karasz FE, MacKnight WJ (1990) Macromolecules 23: 2139
102. Grobelny J, Rice DM, Karasz FE, MacKnight WJ (1990) Polym Commun 31: 86
103. Zhang X, Wang Y (1989) Polymer 30: 1867

# Two-Dimensional Exchange NMR Spectroscopy in Polymer Research

**H. W. Beckham* and H. W. Spiess**

Max-Planck-Institut für Polymerforschung, Postfach 3148, D-55021 Mainz, FRG

**Table of Contents**



---

*Permanent address: School of Textile and Fiber Engineering, Georgia Institute of Technology, Atlanta, Georgia 30332-0295

NMR Basic Principles and Progress, Vol. 32

In recent years the concept of two-dimensional exchange NMR spectroscopy, which is well-established as a tool for structural elucidation of biopolymers in solution, has successfully been applied to solid synthetic polymers. This review introduces the methodology of 2D exchange NMR in solids and illustrates its power by a number of experimental examples. Structural aspects include the packing of amorphous and frozen liquid-crystalline polymers as well as domain sizes in polymer blends. The unique information about slow molecular dynamics provided by 2D exchange NMR is demonstrated for the chain motion of polypropylene in both its crystalline and amorphous modifications. It is shown that 2D exchange NMR is also able to detect chain diffusion between crystalline and amorphous regions. As an example of new polymeric materials, the rotational motion of the discotic units in hairy rod macromolecules is presented. It is shown that the side-group motion in glassy poly(methyl methacrylate) is a rather complex process involving a coupling to a rocking motion of the main chain. Finally it is demonstrated that 2D exchange NMR can also be applied to determine the orientational distribution function in partially ordered polymers. Here a biaxially stretched film of poly(ethylene terephthalate) provides an illustrative example.

# 1 Introduction

No other class of materials is used for such a diverse variety of applications as synthetic polymers. The extensive applicability of polymeric materials arises from the ability to tailor bulk properties by manipulation of the structure, order, and dynamics of the macromolecular chains and their superstructures. The result is a tremendous range of properties exhibited by these materials [1, 2]. Techniques have always been sought which lend insight into these three measures of the polymer solid state. Due primarily to its unprecedented structural specificity, solid-state nuclear magnetic resonance (NMR) has emerged as a particularly versatile method for obtaining the information needed for characterization and investigation of solid polymers. The main advantage of solid-state NMR is the fact that it is based on the use of a completely innocuous but definitive label inherently present, most of the time, within the polymer.

By its very nature, NMR spectroscopy on solid polymers contains more information than can be accessed easily. In contrast to liquid-state NMR, a number of solid-state interactions tend to broaden the spectra rendering them largely indecipherable. In fact, the historical development of high-resolution solids NMR has concentrated on techniques to decrease this spectral broadening by averaging or eliminating such solid-state interactions [3]. However, because interactions such as dipolar coupling and chemical-shift anisotropy contain information on the nature of the solid state itself, a primary objective guiding the recent development of new NMR techniques has been to devise experiments in which this information can be collected and presented in a digestible form. For example, the introduction of additional spectral dimensions [4] (e.g., 2D and 3D) has allowed the presentation of such important parameters as chemical shift, a structure indicator, on one axis, and chemical-shift anisotropy, an indicator of order and dynamics, on the second axis thereby providing a 2D correlation plot of these important molecular characteristics [5].

A virtual explosion of work has recently accompanied solid-state NMR investigations of solid polymers as manifested by the many review articles [6–8], chapters [9, 10], and dedicated books [11–14] that have been published in the past few years, or are about to appear in print. More important, due to the continuing development of methods and techniques, information is becoming accessible through some of the new two-dimensional (2D) solid-state NMR experiments that is simply not available in the same direct manner from any other technique. In particular, 2D exchange NMR has proved to be quite generally applicable for revealing molecular origins of many polymer material properties. Following an introduction of the basics of 2D exchange NMR, examples will be given of its application to investigating the three most commonly named descriptors of the polymer solid state: structure, dynamics, and order. Those 2D NMR experiments not directly yielding information related to the material properties of solid polymers will not be discussed. Cases in which it is necessary, and not just possible, to do a 2D experiment will be highlighted. The basics will be developed and presented in as general a manner as possible; it is emphasized that the same elementary experiment is used to lend insight into the structure, order, and dynamics of solid polymers. Only the specifics of the experiments result in the extraction of the different information. The following review should serve as a guide for polymer scientists interested in discovering what 2D NMR can bring to their research.

## 2 Two-Dimensional NMR

Much of the success of 2D NMR studies has depended on the anisotropic nature of nuclear interactions in the solid state [3]. Thus, not only is the innocuous label already present in most solid polymers, but this label also possesses inherent directionality. When the geometrical connection to the molecular framework is established, a direct link exists between the NMR signal and the molecular orientation [13]. This information can then be exploited to measure the bulk order of the molecules. Similarly, by correlating this molecular orientation after a waiting time with a second orientation of the same moiety, detailed information about geometries of molecular reorientations (i.e., molecular dynamics) are obtained.

Several nuclei have been successfully integrated into 2D NMR investigations of polymer structure, order and dynamics. However, two nuclei have been especially employed in such studies: the isotopes $^{13}C$ and $^{2}H$. The $^{13}C$ nucleus is present at 1.1% natural abundance in all organic compounds and therefore is a relatively isolated label. This is an important characteristic to consider for dipolar nuclei since spin diffusion [15] causes signal averaging across the entire sample if the spins are too concentrated, thereby preventing conclusions from being drawn about different parts of the polymer structure [13]. On the other hand, spin diffusion can be used to give information about the relative proximity of various structural components, and even the relative orientation of such neighboring species. This will be described in Sect. 3.

Whereas the naturally occurring $^{13}C$ spins typically provide sufficient NMR signal, the $^{2}H$ nucleus must be artificially incorporated into the structure of a polymer by isotopic substitution. This is the major drawback as well as the major advantage of NMR of $^{2}H$-labeled polymers: a drawback because isotopic labeling is an expensive and sometimes tedious undertaking. Nevertheless, it often provides considerable benefits because different structural components of the polymer may be labeled selectively which allow for independent determinations of dynamics and order of different parts of the polymer [8, 11]. The $^{13}C$ nucleus may also exhibit the structural specificity required for these types of measurements as long as the overlap of chemical shifts is not too severe for chemically distinct carbons in the same polymer. Although this requirement is only met in the simplest polymers, ingenious schemes have been devised for the separation of the anisotropic part of the chemical shifts, or similarly the spinning sideband patterns, according to the respective isotropic chemical shifts [5]. Thus increasingly complex structures are being completely characterized without the need for sometimes laborious isotopic labeling preparations.

## 2.1 Angular-Dependent Spin Interactions

The orientation-dependent NMR frequency is a tensor quantity and may be described by the following equation [3]:

$$\omega(\theta, \phi) = \omega_L + \delta/2(3\cos^2\theta - 1 - \eta\sin^2\theta\cos 2\phi)$$

Deviations from the Larmor frequency at the isotropic chemical shift ($\omega_L$) are given by the additional angular-dependent term weighted by $\delta$, which represents the strength of the anisotropic NMR interaction. The anisotropic interaction may be defined in terms of principal values in a principal axes system (in which the tensor is diagonal). The polar coordinates, $\theta$ and $\phi$, relate the orientation of the principal axes system (PAS) of the anisotropic-interaction tensor to the external magnetic field. The presence of all possible tensor orientations in equal proportions describes a powder lineshape. The asymmetry parameter $\eta$ $(0 < \eta < 1)$ is the fractional deviation from axial symmetry of the anisotropic interaction. Axially symmetric NMR interactions ($\eta = 0$) reduce the chemical shift to an angular dependence described by $(3\cos^2\theta - 1)$. In this case, the chemical-shift tensor (CST) possesses a unique axis.

The dominating anisotropic interaction for the $^2H$ nucleus is the coupling of the nuclear quadrupole moment with the electric field gradient of the C—$^2$H bond. Because the $^2H$ nucleus is spin $I = 1$, the resulting lineshape for an isotropic rigid solid is the familiar Pake doublet (cf. Fig. 1) [16]. The $z$ axis (unique axis) of the quadrupolar-interaction tensor is oriented along the C—$^2$H bond, a characteristic that makes $^2H$ NMR especially useful for studies of molecular segmental orientations and reorientations. For the heteronuclear-decoupled $^{13}C$ nucleus, the anisotropic magnetic shielding by the electrons at the nucleus gives rise to the chemical-shift dispersion (chemical-shift anisotropy, or CSA). The $^{13}C$—$^1H$ dipolar coupling is also anisotropic and has been quite extensively employed for investigations of solid-state polymer dynamics [17–21]. However, this interaction has so far not been directly exploited in 2D exchange sequences for polymer studies as much as the chemical-shift anisotropy. On the other hand, the $^{13}C$—$^1H$ dipolar coupling has been used in an indirect manner due to its known connection with the molecular geometry: The unique axis is along the $^{13}C$—$^1H$ bond direction [3]. In a 2D experiment, the $^{13}C$—$^1H$ dipolar coupling can be correlated with the CSA lineshape to determine the orientations of the principal values of chemical-shift tensors in the molecular frame [13]. This is an important parameter to be characterized since it establishes the relationship between the spectral frequency and the actual orientation of the molecular segment [22–24]. It is the CSA that has been primarily exploited

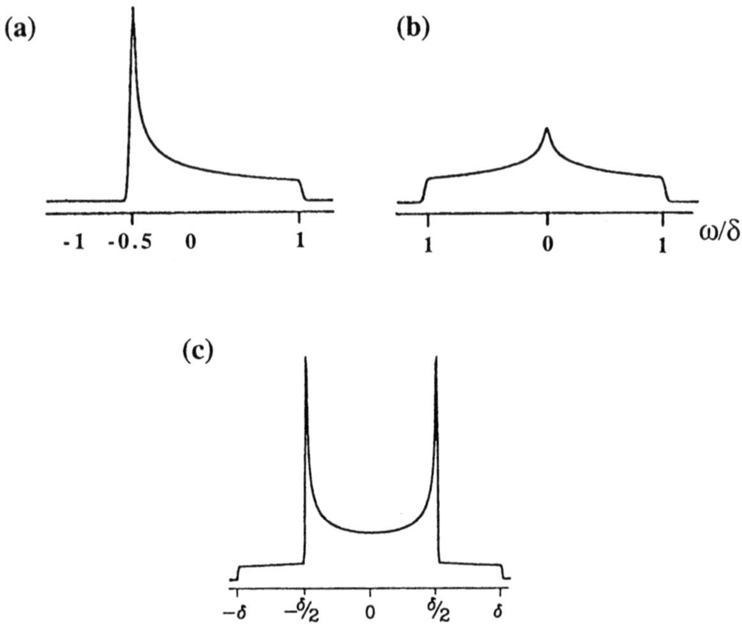


**Fig. 1a–c.** Orientation-dependent NMR lineshapes for powders (i.e., all orientations present with equal probabilities): **a** chemical-shift anisotropy, $\eta = 0$; **b** chemical-shift anisotropy, $\eta = 1$; **c** $^{2}H$ quadrupolar coupling or spin-1/2 dipolar-pair coupling, $\eta = 0$ (Pake pattern) [3, 13]

in $^{13}C$ 2D exchange NMR on solid polymers, while the $^{13}C$—$^{1}H$ dipolar coupling is normally removed by high-power proton decoupling. The $^{13}C$ nucleus is spin $I = 1/2$ and gives rise to an anisotropic powder lineshape defined simply by the principal values of the chemical-shift tensor (in the absence of molecular motions of rates with Fourier components in the range of the inverse linewidth). The CSA lineshape is characteristic for each nucleus type and often reflects the local bonding symmetries [25]. This lineshape, along with the quadrupolar-coupled $^{2}H$ lineshape (the $^{13}C$—$^{1}H$ dipolar-coupled lineshape has the same form), is illustrated in Fig. 1.

For the anisotropic $^{13}C$ chemical shift, sufficiently rapid sample spinning about $\theta = 54.7°$ (with respect to the external magnetic field) effectively averages the orientation dependence to zero and gives the isotropic chemical shift [3]. This technique of magic angle spinning (MAS) has been employed in conjunction with high-power proton decoupling and cross-polarization for obtaining high-resolution $^{13}C$ spectra of rigid polymer solids for quite some time [26, 27]. In many 2D experiments, fast MAS is naturally employed to provide the isotropic chemical shifts in either one or two dimensions. The nuclear orientation information may be partly retained, and thus exploited, by spinning the sample at a rate slower than the chemical-shift anisotropy. Under such slow-MAS conditions, envelopes of sidebands spaced at multiples of the spinning speed appear which roughly silhouette the static continuous CSA lineshapes [11].

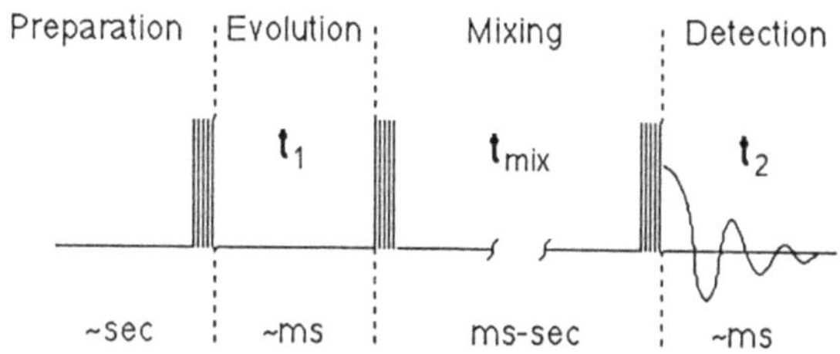


**Fig. 2.** Generalized timing diagram of the three-pulse sequence used in 2D exchange spectroscopy. Relative durations of each period are indicated. The preparation period typically includes a relaxation time for the magnetization to return to equilibrium. Molecular-segment orientations, contained in the nuclear frequencies $\omega_1$ and $\omega_2$, are measured during the evolution ($t_1$) and detection ($t_2$) periods. Orientation changes, by any mechanism, are probed during the mixing period ($t_{mix}$) [28]

These NMR interactions may be manipulated in 2D exchange spectroscopy by the generalized three-pulse sequence shown in Fig. 2 [28]. After the initial pulse, the evolution of the nuclear magnetization during $t_1$ effectively labels the beginning molecular orientation. This is stored by the second pulse. A variable mixing time follows during which the NMR frequency (reflecting the initial molecular orientation) may exchange to reflect a final molecular orientation. This NMR frequency is subsequently recorded during the detection time following the third pulse [4]. The times needed for the evolution ($t_1$) and detection ($t_2$) periods are usually much shorter than the delay introduced to probe exchange during the mixing time ($t_{mix}$). By simply repeating the pulse sequence with incremental increases of the evolution time, a two-dimensional "snapshot" of the nuclear exchange event may be generated for a single mixing time by double Fourier transformation across the evolution and detection periods: $S(\omega_1, \omega_2) = FT_1\{FT_2[s(t_1, t_2)]\}$ [4]. The result is an exchange pattern that characterizes the correlation of the anisotropic-interaction tensors before and after the mixing time. The actual mechanism of exchange governs the type of information extracted from the experiment [13]. As will be shown in detail below, if the NMR frequency exchanges via a process called spin diffusion between neighboring nuclei of fixed locations, then information on relative orientations, and thus the localized molecular structure, is obtained. If the NMR frequency exchanges simply due to molecular reorientation, then information on the molecular dynamics is found. If the NMR frequency exchanges after deliberate reorientation or flipping of thc entire sample with respect to the external magnetic field, then information on the molecular order is available.

If an exchange event does not occur during the mixing time, then $\omega_1 = \omega_2$ and the 2D spectrum consists of the anisotropic lineshape along the diagonal of an $S$ versus $(\omega_1, \omega_2)$ plot. If exchange occurs, by any mechanism, then the 2D spectrum contains off-diagonal signal intensity, the shape of which is determined by the specifics of the exchange process. For discrete exchange processes such as spin diffusion between nuclei existing in only two relative

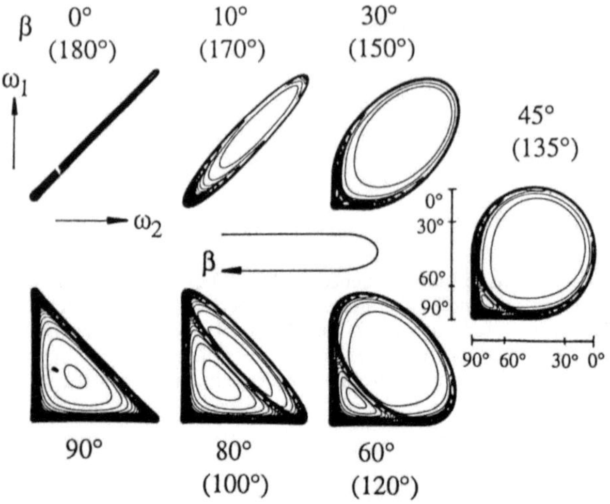


**Fig. 3.** Exchange powder patterns for $\eta = 0$ interaction tensors correlated by discrete single-axis rotations through $\beta$. These 2D patterns are typical for spin-1/2 anisotropic-chemical-shift lineshapes. The $^2H$ quadrupolar-coupling and dipolar-pair-coupling patterns are similar, and are produced by superimposing these patterns with their reflections about ($\omega_1 = 0, \omega_2 = 0$). Due to the magnetic-field symmetry axis, patterns for $\beta$ and $180° - \beta$ are indistinguishable [13]

orientations (structure), a two-site molecular-segment jump (dynamics), or any macroscopic sample flip (order), the 2D exchange patterns will characteristically reflect the relative orientations of the two tensors correlated for that particular experiment [13]. For the spin-diffusion example, since the correlated tensors correspond to two different nuclei, then the $\delta$ and $\eta$ must be sufficiently similar for this statement to hold (often the spin-diffusion exchange occurs between two chemically identical nuclei). Typical exchange powder patterns for discrete processes are shown in Fig. 3 for the $\eta = 0$ CSA lineshape depicted in Fig. 1. The most striking spectral features apparent in the patterns of Fig. 3 are the ellipses, and the straight ridges parallel to the frequency axes [13]. From these elliptical ridges, the relative-orientational (structure) or reorientational (dynamics) angles can be directly calculated [the flip angle (order) can also be directly calculated but is a fixed experimental parameter] by relating the major and minor axes of the ellipse: $|\tan \theta| =$ *perpendicular axis length/diagonal axis length* [29]. Underlying the generality of this technique, in certain instances the 2D exchange patterns will be identical for completely different exchange mechanisms: usually for axially symmetric tensors and when the transformation between the two correlated tensors is reflected in a single reorientational angle. For increasingly indiscrete exchange processes, these well-defined patterns will be increasingly broadened around their maxima. This would arise from spin diffusion between nuclei existing with a distribution of local relative orientations (structure) or a distribution of molecular-segment reorientational jump angles (dynamics). Angular distributions correlating the anisotropic-interaction tensors before and after $t_{mix}$ may be reconstructed directly from the 2D powder spectra [13, 30].

Thus, the analysis of the 2D exchange spectra is mostly independent of the type of information being sought. The experiment is valuable because angular information, as embodied in the equation shown above for the orientation dependence of the NMR frequency, may be extracted directly from the spectra. Since the molecular orientations are measured twice (in the evolution and detection periods), a 2D spectrum simply reflects the distribution of principal axes systems with initial orientations $(\theta_1, \phi_1)$ whose final orientations are $(\theta_2, \phi_2)$. Stated more precisely, a 2D exchange spectrum is a visual representation of a "two-time" distribution function [31]. The actual property characterized by this two-time distribution function depends on the specifics of the 2D exchange experiment and the system under investigation.

## 3 Structure: Exchange by Spin Diffusion

The area of nanostructured materials is receiving an increasing amount of attention from scientists of many disciplines [32]. When the heterogeneities of a material can be controlled on this length scale, a relatively large volume of the material consists of phase boundaries or interface. Thus, novel properties can result. Polymers promise to play an important part in the development of these materials. Techniques to describe structures with characteristic dimensions in the nanometer range have in fact been developed: NMR spin-diffusion measurements are able to distinguish "domains" or more generally, heterogeneities with smallest typical dimensions between 0.5 nm and 200 nm [13]. Nuclear magnetization, initially selected for one phase, is allowed to diffuse during a variable mixing period into other phases. This diffusion occurs via successive energy-conserving spin flips of homonuclear dipolar-coupled nuclei [15]. Thus, the spin-diffusion constants are directly related to the dipolar couplings even if these are reduced by molecular motion [33, 34].

The technique depends upon first selecting the nuclear magnetization for one of the domains, and then monitoring the diffusion of this magnetization out of these regions. In the framework of the generalized 2D pulse sequence of Fig. 2, the selection is done before or during the evolution period, the spin diffusion occurs during the mixing period, and the distribution of diffused magnetization among the various nuclei of the sample is measured in the detection period. The selection can be achieved based on differences in molecular mobility [35, 36], or differences in molecular structure as reflected in the chemical shift (precession frequency) [37, 38]. During the mixing time, spin-lattice relaxation occurs in addition to spin diffusion; this must be accounted for in the data analysis. The spin-diffusion time must be significantly shorter than the spin-lattice relaxation time ($T_1$) of the nucleus under observation for signal to be measured. For protons, with their nearly 100% isotopic abundance and strong dipolar couplings, much spin-diffusion work has been reported [13]. On the other hand, due primarily to the low natural abundance of 1.1%, $^{13}C$ spin diffusion has only been reported in a few materials (without isotopic enrichment). In polyethylene, for instance, the extremely long $T_1$'s of the crystalline regions ($\sim$ 700 seconds) made this possible [39]. Normally, isotopic enrichment of such low abundance nuclei is needed to sufficiently speed up the spin diffusion. However, new experiments have appeared which effectively speed up the spin diffusion by clever manipulation of nuclear spin interactions. Called rotor-driven [40] and radio-frequency-driven [41] spin diffusion, these techniques enable spin-diffusion studies on $^{13}C$ at natural abundance. Thus, advantage may be taken of the superior resolution of $^{13}C$ spectra in such investigations.

## 3.1 Local Relative Orientation

As long as the chemical-shift anisotropy is preserved, and resolvable, exchange by spin diffusion gives relative orientations of neighboring nuclei within a solid [42]. This information is analogous to that obtained by wide-angle X-ray diffraction except that the existence of regular molecular spacings is not required to extract useful data. Thus, spin diffusion promises to be particularly useful for providing short-range geometric information in amorphous materials. One of the first examples was the detection of intermolecular natural-abundance $^{13}C$ spin exchange in a static powder of semicrystalline polyethylene (PE) [39]. Since the experiment was conducted on the static powder, the single resonance of polyethylene displayed the typical CSA lineshape for methylene carbons [25]. This is shown in Fig. 4a along the diagonal of the 2D exchange spectrum for $t_{mix} = 10$ ms, a mixing time not adequately long for exchange to occur as evidenced by the absence of off-diagonal signal intensity (i.e., $\omega_1 = \omega_2$). The 2D exchange spectrum for a much longer mixing time (10 s) is presented in Figure 4b, and clearly indicates exchange only between the $\omega_{11}$ and $\omega_{22}$ parts of the lineshape (as labeled in the figure). During this 10-s mixing time, the $^{13}C$ magnetization in the amorphous components of the polymer relaxes via spin-lattice processes so that the spectrum represents only the crystalline portions of polyethylene. Since the direction corresponding to the $\omega_{33}$ part of the lineshape does not change, the exchange ridges indicate a 90° rotation of the chemical-shift tensor around this axis. This transformation was simulated and is depicted in Fig. 4c. From this point, assuming that the exchange intensity arises only from spin diffusion, a priori knowledge of the $^{13}C$ chemical-shift tensor orientation within the molecular frame could be used to determine the relative orientation of adjacent molecular segments. For the case of polyethylene, the $^{13}C$ chemical-shift tensor orientation as well as the crystalline structure are already known and were used to interpret the spin-diffusion data. As shown in Fig. 4d, the $\omega_{33}$ part of the lineshape corresponds to the chain axis, and the 2D spin-diffusion data confirm the local chain structure within the crystallites of polyethylene: Adjacent chains are in fact rotated by 90° with respect to each other. Thus, this example nicely demonstrated the feasibility of the experiment. However, it provided no new insight into the PE structure.

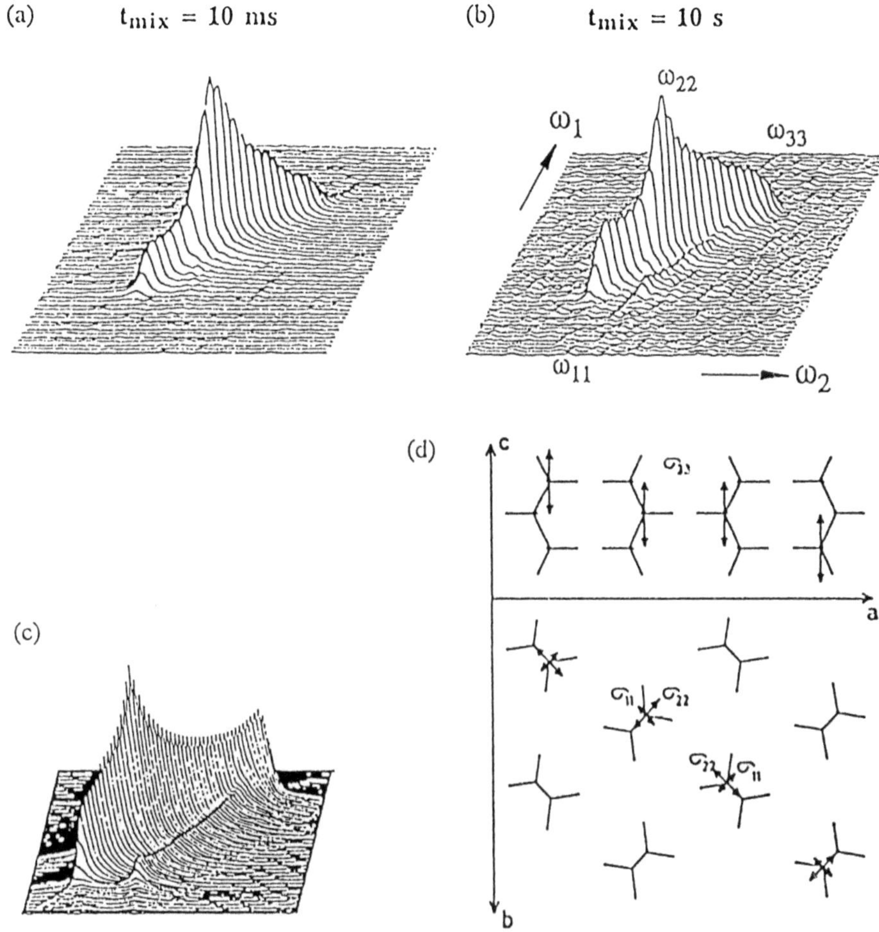


**Fig. 4a–d.** Local relative crystalline-phase chain structure of semicrystalline polyethylene. 2D $^{13}C$ spin-diffusion exchange spectra of the crystalline-phase methylene resonance (room temperature): **a** $t_{mix} = 10$ ms, **b** $t_{mix} = 10$ s, **c** simulation. **d** Relative local orientation of chains within crystalline structure. Projections on two crystal planes are shown. The orientation of the CST principal values ($\sigma_{11}, \sigma_{22}, \sigma_{33}$) in the molecular framework are indicated for the orientationally different methylene groups in the unit cell. (Adapted with permission from Edzes et al. [39])

Amorphous polymers do not usually consist of the same kind of specific relative arrangements of adjacent chains. Thus, the 2D spin-diffusion exchange spectrum is not expected to exhibit any patterns characteristic of geometrically specific exchange paths. Instead, the exchange pattern resembles that shown in Fig. 5 for poly(methyl methacrylate) (PMMA) fully $^{13}C$-labeled at the ester-methyl group [43]. Upon increasing the mixing time from 0.1 s to 0.5 s, off-diagonal exchange is observed between all values of the chemical-shift tensor: Hence, the ester-methyl groups are more or less randomly oriented. For longer mixing times, the exchange intensity simply increases in relation to the diagonal.

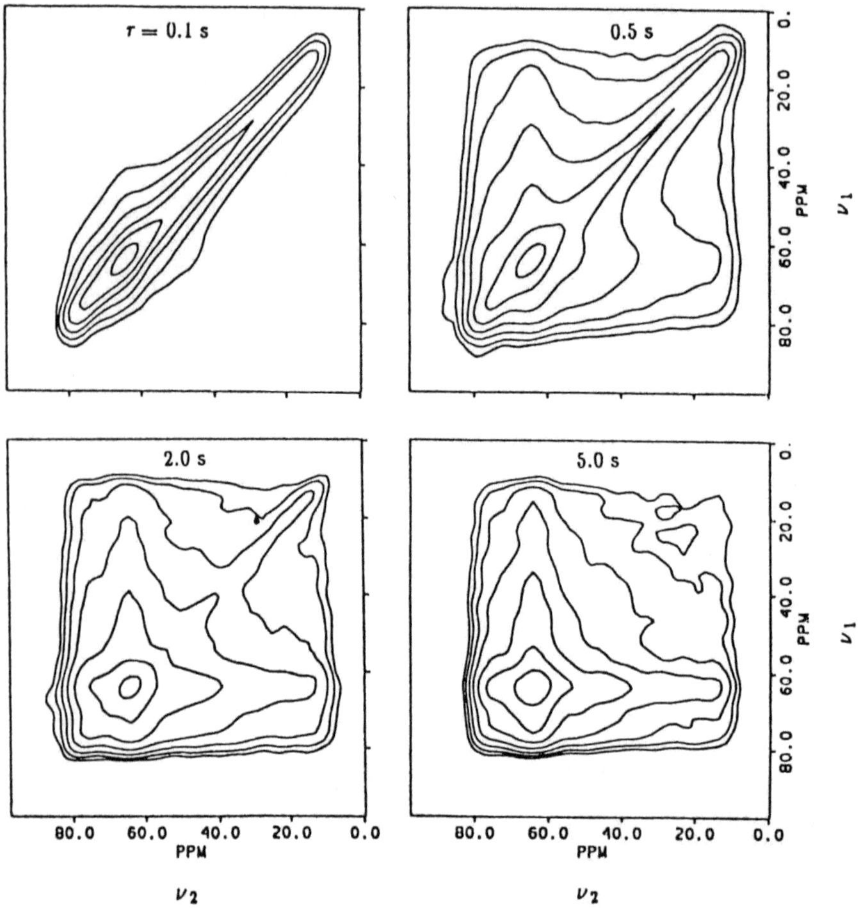


**Fig. 5.** 2D $^{13}C$ spin-diffusion exchange NMR spectra of atactic poly(methyl methacrylate) fully $^{13}C$-labeled at the ester-methyl position (room temperature). Mixing (i.e., spin-diffusion) times are indicated above each contour plot. With increasing $t_{mix}$, pattern development toward one with off-diagonal intensity at all possible combinations of $\omega_1$ and $\omega_2$ is diagnostic for materials containing distributions of all possible local relative molecular-segment orientations (i.e., amorphous). (Reprinted with permission from Tycko et al. [43])

It was specifically stated that molecular motion was "negligible" in PMMA at room temperature so all off-diagonal exchange intensity could be attributed to spin diffusion. Recently, it was found that molecular motion is not negligible, but is of such a type that it would only contribute some exchange intensity close to the diagonal for this ester-methyl $^{13}C$-labeled PMMA [111]. This work will be presented in detail in Sect. 4. It is mentioned here to stress the importance of distinguishing exchange due to molecular motion, and exchange due to spin diffusion. As presented earlier, isotopic enrichment of dilute nuclei such as $^{13}C$ is nearly always required to make the spin-diffusion experiment practical. This could also serve to accelerate the spin diffusion to a time regime where molecular motions are insignificant. Also, molecular-dynamics exchange exhibits a much stronger temperature dependence as compared to spin-diffusion exchange. Since most dynamic processes are rendered very slow at sufficiently low temperatures (which are sample-dependent), spin-diffusion experiments can simply be done below these characteristic temperatures. Conversely, when molecular-dynamics exchange is measured, spin-diffusion exchange should be avoided. The success of the $^{13}C$ nucleus in dynamics studies is due in part to its low natural abundance, and therefore very low spin-diffusion rate [13]. Similarly, one of the major advantages of employing the quadrupolar $^{2}H$ nucleus for molecular-dynamics investigations is its small dipole moment, which results in low spin-diffusion rates even for fully deuterated materials [44].

Unless the orientational correlation lengths are longer than the spin-lattice relaxation times, long mixing times will always lead to exchange patterns like that of Fig. 5 as all CSA tensor orientations are detected (i.e., a powder pattern). On the other hand, if there exists some short-range orientational correlation, this may be detected at short mixing times in terms of a distinct exchange pattern. As the mixing time is increased, larger length scales are probed and will be reflected in the exchange patterns. Thus, the mixing-time dependence of the 2D exchange patterns may provide molecular orientational correlation functions. This information would be extremely useful for seemingly isotropic materials in which orientational correlations exist over nm length scales [43].

The technique was recently applied to study the microscopic arrangement of mesogens in macroscopically ordered side-chain liquid-crystal polymers [45]. These polymers have flexible backbones to which rigid mesogens are connected as side groups through flexible spacer groups. They can be aligned by various methods and display increasing degrees of order in going from the backbone to the mesogenic side groups. The problem is sketched in Fig. 6a, which displays two possible microscopic arrangements of the mesogenic side group within the layers of a smectic polyacrylate (structure shown in Fig. 6b). The macroscopic order of the material can be measured by NMR techniques that exploit the orientation dependence of either CSA tensors, or $^{2}H$ field-gradient tensors (see Sect. 5). The measured bulk order parameters may be derived from two possible short-range relative orientations of the mesogenic groups: a statistical arrangement, or a gradually changing organization. In the former case, the short-range order is the same as the measured bulk order, while in the latter case, the short-range order of ensembles of neighboring mesogens would be higher. This distinction was made by monitoring the spin diffusion between mesogens $^{13}C$-labeled (95%) in the carboxyl group. The 2D spin-diffusion exchange data are shown in Fig. 6b. A short mixing time (300 ms) probes immediate neighbors. A long mixing time (3 s) probes the same region but with a three-times larger diameter, thereby depicting the measured bulk order. Already for the short mixing time, the exchange pattern is virtually identical to that recorded for the long mixing time, the only difference being increased off-diagonal exchange intensity. Thus, the microscopic order is not higher than the macroscopic order.

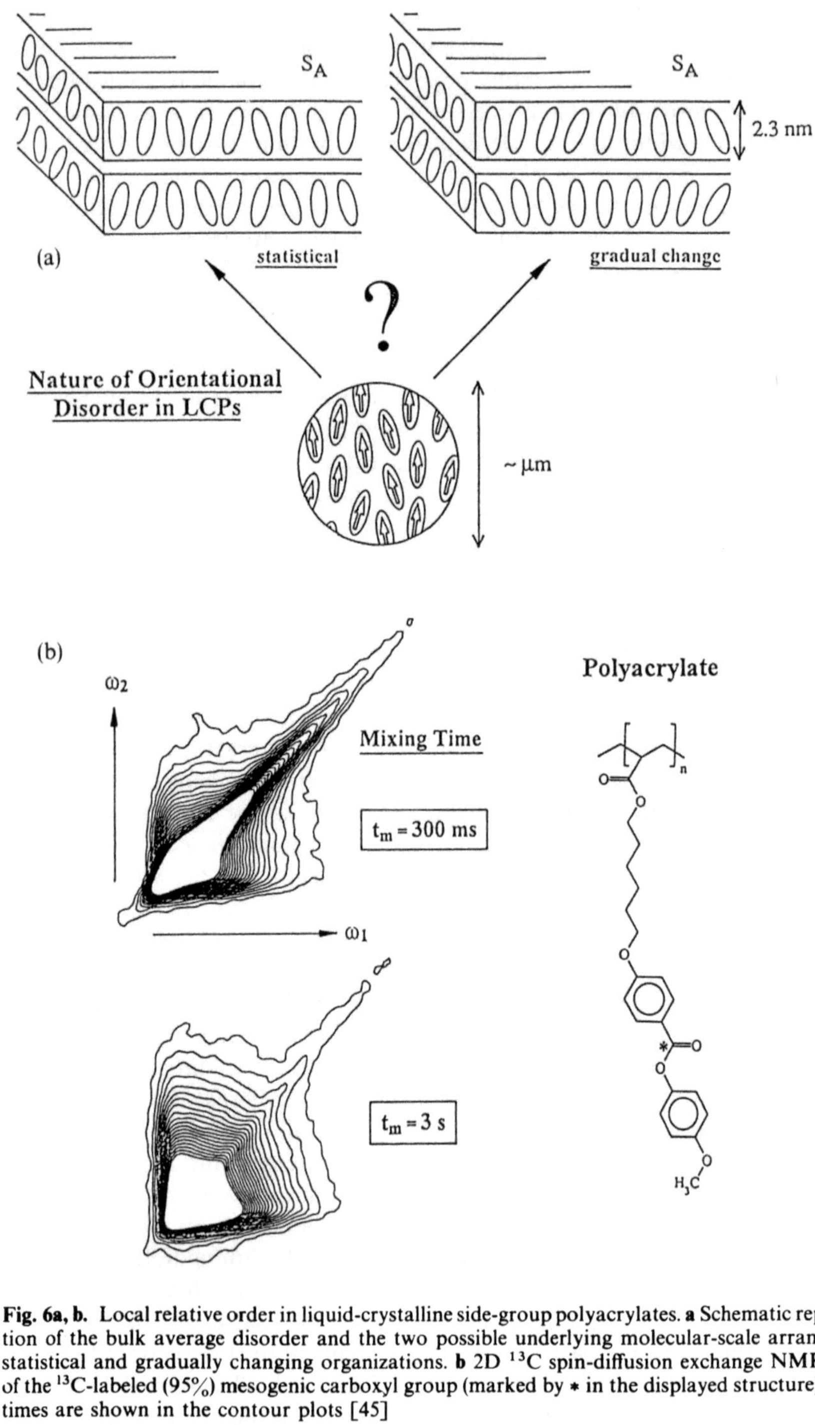


**Fig. 6a, b.** Local relative order in liquid-crystalline side-group polyacrylates. **a** Schematic representation of the bulk average disorder and the two possible underlying molecular-scale arrangements: statistical and gradually changing organizations. **b** 2D $^{13}C$ spin-diffusion exchange NMR spectra of the $^{13}C$-labeled (95%) mesogenic carboxyl group (marked by * in the displayed structure). Mixing times are shown in the contour plots [45]

Therefore, the local arrangement of the mesogenic side groups must be described as statistical [45].

## 3.2 Molecular Proximity

Notwithstanding the previous examples, most of the spin-diffusion work reported so far has monitored the proximity of molecular segments or domains, from which specific molecular interactions (e.g., mixing on a molecular level) have been determined. Because higher spectral resolution allows more detailed structural assignments, these types of studies have primarily monitored spin diffusion among nuclei via their respective isotropic chemical shifts. For concentrated nuclei such as $^{1}H$, homonuclear dipolar couplings must be removed by an appropriate pulse sequence such as MREV-8 in order to resolve the isotropic chemical shifts [46, 47]. For dilute nuclei such as $^{13}C$, this resolution can be achieved by magic-angle spinning to eliminate CSA broadening, in conjunction with high-power decoupling to eliminate the heteronuclear dipolar coupling to the protons [48]. On the other hand, molecular reorientational rates of about $10^{7}\,s^{-1}$ and faster preclude the need for MAS [49]. Another possibility for removing CSA overlap is to orient the sample so that only part of the CSA tensor contributes to the static lineshape. This was demonstrated for $^{15}N$ spin exchange on an oriented protein ($^{15}N$-labeled) in which the partially resolved static lineshapes revealed some connectivity relationships through a 2D exchange pattern [50]. Normally, however, experiments to explore molecular proximities employ MAS. For example, the miscibility of an amorphous blend of poly(ethylene terephthalate) (PET) and the polycarbonate of bisphenol-A (PC) was investigated in which one of the two components of the blend was $^{13}C$-labeled [51]. Using a mixing time of 0.4 seconds, crosspeaks between the enriched methylene carbons of PET and the peaks due to the PC backbone clearly indicated site-specific interpolymer spin diffusion. Data analyses based on crosspeak intensities and theoretical models indicated that the blend was homogeneously mixed on a length scale of 4.5 to 6 Å.

## 3.3 Domain Sizes

In these studies it was suggested that 2D NMR spectroscopy was "required" for mapping out the different spin-diffusion pathways. This is definitely true for systems with many different possible spin-diffusion paths in which it is not possible to select unambiguously the source or beginning magnetization either experimentally based on molecular mobility or chemical shift, or chemically by specific isotopic enrichment. However, when the selection can be clearly made, then it is sufficient, and even desirable to measure one-dimensional spectra and monitor the spin diffusion as a function of mixing time via peak intensity changes. This technique has been thoroughly developed and successfully applied, primarily with $^1H$ spin diffusion to determine domain sizes of semicrystalline bulk homopolymers [52], fibers [53], and copolymers [54], as well as various polymer blends and amorphous diblock structures [55].

Analogous information is provided by a heteronuclear 2D MAS experiment—$^1H$ wideline spectra are separated along one axis according to their respective $^{13}C$ chemical shifts—in which $^1H$ spin diffusion is monitored via the $^1H$ lineshapes at each $^{13}C$ chemical shift representing a different structural component. The 2D WIdeline SEparation (WISE) experiment [56, 57] is particularly facile to implement and provides direct information on relative structural mobilities in dynamically heterogeneous materials. It takes advantage of the dynamics information content of $^1H$ wideline spectra. Rigid structural units produce broad lines which are narrowed by molecular motions on the timescale of tens of kHz [15]. For materials containing both rigid and mobile components, corresponding 1D $^1H$ wideline spectra consist of a narrow line superimposed over a broad line. In the 2D WISE experiment, cross-polarization of the $^1H$ magnetization to proximate $^{13}C$ spins transfers this mobility information to a larger chemical-shift dispersion ($^{13}C$), thus affording resolution and connection with different structural components. When a mixing time is inserted into the pulse sequence, spin diffusion between these separate components may be monitored. By varying the length of the mixing time, $^1H$ spin diffusion is observed indirectly. While in normal exchange-type experiments, this is marked by the presence of off-diagonal crosspeaks, in the 2D WISE experiment spin diffusion is followed via the shape of the $^1H$ widelines for each $^{13}C$ peak. That is, equilibrium is reached when a $t_{mix}$ has been identified in which all $^1H$ lines have the same shape (i.e., with narrow and broad components). With knowledge of the spin-diffusion coefficients, the relative sizes of the domains may be estimated [13, 55].

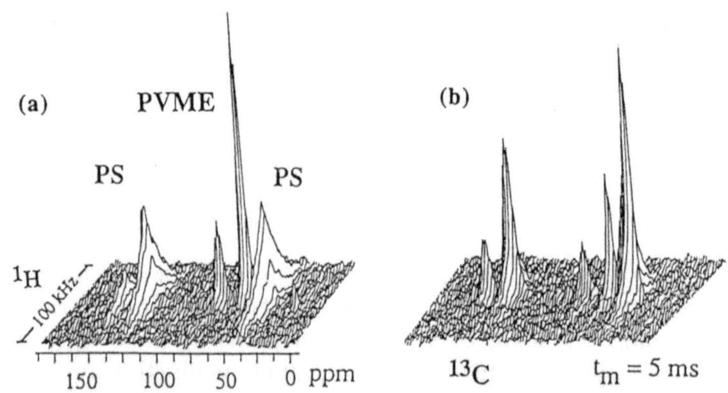


**Fig. 7a, b.** 2D WISE spectra of a 50:50 blend of polystyrene (PS) and poly(vinyl methyl ether) (PVME) cast from toluene: **a** no mixing time, and **b** $t_{mix} = 5$ ms. For the spectrum with no mixing time, the narrow $^1H$ lines of the PVME reflect its higher segmental mobility as compared to the PS whose $^1H$ lines are significantly broader. These $^1H$ linewidths have equilibrated already after $t_{mix} = 5$ ms, which establishes an upper limit for the size of the heterogeneities [56]

An example is presented in Fig. 7, which displays the 2D WISE spectra as a function of mixing time for a 50:50 blend of polystyrene (PS) and poly(vinyl methyl ether) (PVME) cast from toluene [56]. With no mixing time, Fig. 7a clearly reveals that this blend possesses areas of varying mobility; thus, the blend is not homogeneous. This is in contrast to findings by such classical measurements as differential scanning calorimetry (DSC), which indicates a single glass-transition temperature ($T_g$). From the $^1H$ lineshapes in the spectrum of Fig. 7a, it can be seen that the PVME is mobile while the PS is considerably more rigid. After a 5-ms mixing time (Fig. 7b), spin diffusion has equilibrated the magnetization across the sample as evidenced by identical $^1H$ lineshapes for all $^{13}C$ peaks. Utilizing diffusion coefficients determined in systems whose domain sizes were also characterized by transmission electron microscopy [58], and whose $^1H$ linewidths are similar, domain sizes were estimated to be $3.5 \pm 1.5$ nm. Thus, the PS/PVME 50:50 blend cast from toluene must be more properly described as nanoheterogeneous. This type of detailed materials characterization on these length scales is one of the main advantages of NMR spin-diffusion studies.

## 4 Dynamics: Exchange by Molecular Motion

Although the bulk material behavior of solid polymers ultimately depends on a combination of many factors such as morphology and degree of preferred orientation, the physical characteristics of pure polymers can usually be directly linked to the type and timescale of molecular motions. Having characteristic relaxation times ranging from tens of picoseconds to minutes [59], polymers have been synthesized and fabricated for such uses as hydraulic fluids and elastomers (fast relaxation times), to structural components for automobiles and consumer electronics (slow relaxation times). Clearly, a molecular description of the polymer chain dynamics would provide meaningful insight into the nature of these materials, and by comparing the findings with bulk macroscopic properties provide guidance for producing new generations of technologically useful materials.

### *4.1 Background*

The characterization of polymer molecular motion has traditionally been accomplished with dynamic mechanical, dielectric, and wideline NMR spectroscopy as a function of temperature or frequency [60]. "Transitions" appeared in these spectra and were consequently assigned to the motional onset of various parts of either the side chain or main chain of the polymer on the timescale of the corresponding measurement. However, with these traditional relaxation techniques, it is not easy to directly connect spectral transitions with specific molecular motional mechanisms. Assignments of the loss peaks in dynamic mechanical or dielectric loss spectra are usually made indirectly by comparison of data collected on structurally similar polymers in which changes in the relaxation spectra are associatcd with the structural differences of the polymers [61, 62]. Thus, unambiguous descriptions of the polymer dynamics are usually not possible.

Through various experiments, NMR spectroscopy can be used to characterize molecular motions over virtually the entire range of characteristic relaxation times present in polymers [6–15, 63, 64]. In addition, the molecular specificity inherent to NMR methods allows the direct connection of motional information with molecular structure. Relaxation-time measurements have for a long time been used to extract timescale information. Separate spectral peaks for chemically different moieties enable relaxation times to be measured for different parts of the structure, thus revealing a great deal over the traditional techniques. Even if a single peak represents contributions from two different domains (e.g., the amorphous and crystalline regions of semicrystalline polymers), appropriate models can be applied and relaxation times extracted for the separate domains [65, 66].

Relaxation-time studies have been employed to provide relative measures of polymer molecular level dynamics [12, 14]. From these, correlations have been found with such macroscopic physical properties as mechanical behavior and transport properties of gases and liquids through the polymer matrix. Connections have been drawn between molecular mobilities and diffusivities of low-molar-mass diluents and gases in solid polymers: higher diffusivities when molecular motions are characteristically faster [67]. For example, $^{13}$C NMR rotating-frame relaxation rates were used to explain the antiplasticization effect of a low-molar-mass diluent by reduced main-chain motions, and the subsequent plasticization effect by increased main-chain motions in polycarbonate [68].

A great deal of work has been reported in which the mechanical impact strength of a polymer is linked to the molecular-level dynamics [69]. The majority of these efforts noted the frequent coexistence of high impact strength with a significant low-temperature loss peak in the dynamic mechanical spectrum. The contention was that this low-temperature loss peak represented some sort of localized segmental motion that acted to damp the force of the impact through matched spectral densities between the mechanical impact and the molecular motion. This hypothesis was further supported with an NMR relaxation-time study which found a direct correlation between impact strength in glassy polymers and a relaxation-time parameter that represented molecular motions in the same frequency range as the mechanical impact [70]. Prompted by these findings, other researchers have correlated the dynamic storage moduli for a number of amorphous as well as semicrystalline polymers with an NMR relaxation constant characterizing near-static molecular motions [71].

Further information is provided by one-dimensional static lineshape studies [3, 11]. As presented in Sect. 2, solid-state NMR lineshapes are anisotropic and depend on the orientation of the chemical moiety with respect to the external magnetic field. Static lineshapes are changed in sometimes characteristic ways by specific molecular motions that occur on the same timescale as the width of the lineshape. Various motional models such as *n*-site jumps and rotational isotropic diffusion are invoked to explain the relaxation-time and lineshape changes with temperature [72]. With the help of such motional models, the nature or geometry of such motions may be determined. For example, evaluation of $^{31}P$ CSA lineshapes as a function of temperature for phosphate esters in glassy polymers has provided some insight into glass-transition dynamics in polymer/diluent systems [73]. Deuterium [74, 75] and $^{13}C$ static lineshape [76] analyses as a function of temperature first revealed the now well-known 180° flip of the backbone phenyl group in polycarbonate, motion that has been quantitatively related to the low-temperature dynamic-mechanical loss peak [77]. This phenylene flip motion has been used as the basis for computer modeling of the chain dynamics of amorphous polycarbonate [78]. This type of simulation promises to provide valuable information on the cooperative nature of both intrachain and interchain motional behavior, information that is not readily available experimentally, but important for describing macroscopic mechanical properties such as yielding. This 180° phenylene flip motion has since been shown to occur in other phenylene-containing systems such as poly(phenylene sulfide) [79], poly(butylene terephthalate) [80], polysulfones [81], poly(*p*-phenylene terephthalamide) [82], and various liquid-crystalline polymers [83].

Sensitive to motions in the timescale regime from $10^{-7}$ to $10^{-2}$ seconds, relaxation-time measurements and lineshape studies do not directly probe motions that are slower [3]. However, this "ultraslow" regime is particularly important for solid polymers as the range in which main-chain motions occur, motions most likely the primary determinants of macroscopic physical properties [84]. Sensitive to this time regime, 2D solid-state exchange NMR techniques have proven especially useful for revealing these polymer main-chain motions [9, 13]. Since the same timescales are probed in the traditional techniques of dynamic mechanical and dielectric spectroscopy, direct comparisons may be made with data collected with these methods without the need for time/temperature superpositions (as is necessary for the lineshape and relaxation-time data) [11, 14].

Motions too slow to cause 1D lineshape changes are observed in 2D exchange NMR. Described in Sect. 2, the 2D exchange technique measures molecular orientation before, and after a variable mixing time. If the molecular orientation has been changed during the variable mixing time by some dynamic event, then this will appear on the 2D plot as off-diagonal exchange intensity. At a single mixing time, the 2D NMR plot immediately gives information on the geometry of motion active in molecular reorientation without the need for invocation of motional models [13]. Small-angle step-diffusive molecular motion is manifested by a general spreading from the diagonal into the 2D plane. Discrete large-angle jump motions between primary sites will be exhibited as well-defined ridge patterns, examples of which are shown in Fig. 3. One-dimensional lineshape studies provided overwhelming evidence for the existence of the 180° phenylene flip motion [74–76], but this motional model used to interpret the data is in fact not unique. The final proof that the phenylene motion in polycarbonate was a 'flip' between two sites was provided by 2D exchange NMR [85]: The sample was cooled to bring the timescale of the motion into the slow-motion limit, and the appearance of the characteristic elliptical ridge patterns confirmed a rather well-defined jump process.

## *4.2 Comparison of Techniques: Polypropylene*

To demonstrate the utility of 2D exchange NMR in polymer-dynamics research, a collection of studies on a single polymer type, polypropylene, will be presented. In this manner, various exchange experiments may be compared for characterizing amorphous-phase as well as crystalline-phase main-chain motions in solid polymers. Following this section, the flexibility of the techniques for detecting different types of motions will be demonstrated by presenting some recent work on other types of polymers: semicrystalline polyethylene, amorphous poly(methyl methacrylate), and a novel liquid-crystalline polymer.

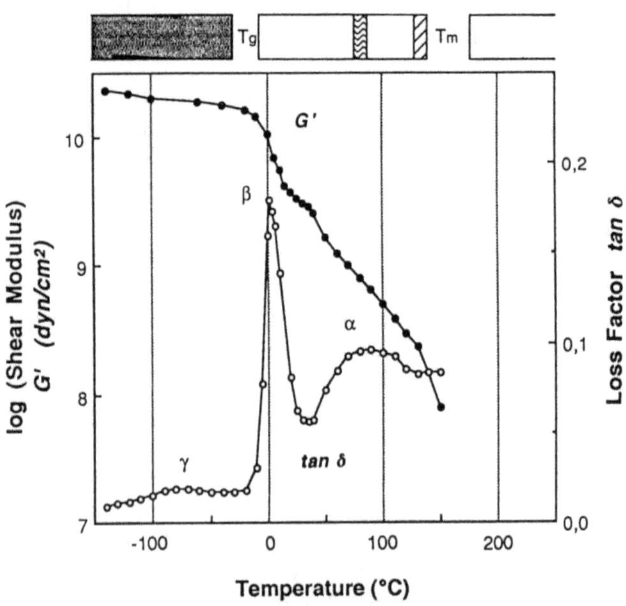


**Fig. 8.** Dynamic mechanical data for isotactic polypropylene, plotted as the loss factor (tan $\delta$) and log of the shear modulus as functions of temperature (adapted with permission from Flocke [86]). Transitions marking the onset of various motional processes ($\alpha, \beta, \gamma$) are labeled in the tan $\delta$ curve. The temperature regions of these transitions, as well as intervening property ranges are summarized in bar-graph form over the top of this plot: ▨ transition to elastically tough, ▨ thermoplastic processing region (adapted with permission from Franck et al. [87])

Polypropylene is a very common polyolefin that is known in its atactic, syndiotactic, and isotactic forms, the latter of which is found in many various commercial applications [1]. In Fig. 8, dynamic mechanical data for isotactic polypropylene are exhibited in the form of the shear modulus and loss factor as a function of temperature [86]. Over the top of this plot is a bar-graph representation of the "transition" temperatures and intervening property ranges of this polymer [87]. This is the type of information that is useful to engineers and designers for material selection purposes. Connection of these macroscopic bulk properties with specific molecular motions is a major impetus for thc 2D solid-state NMR studies. As will be shown, much is now known, but not all. For example, the small low-temperature peak of Fig. 8, labeled $\gamma$, has not yet been unambiguously identified with a specific molecular process, but is considered to involve some type of local motion of restricted amplitude within the amorphous regions. On the other hand, the $\beta$ peak, long associated with the glass transition, and the $\alpha$ peak, long connected with the presence of crystallinity [60], have been identified in more detail as $^{13}C$ and $^{2}H$ 2D exchange on static samples of atactic and isotactic polypropylene have been performed [88, 89]. Spectra are shown in Figs. 9–12 for comparison of both the polymers and the techniques [13]. Assuming the chain dynamics of atactic polypropylene also represent the chain dynamics within the amorphous regions of isotactic polypropylene, then 2D exchange studies of the atactic polymer give insight into the $\beta$ process of thc isotactic polymer.

In Fig. 9a is shown the $^{2}H$ 2D exchange spectrum for atactic polypropylene near the glass transition [89]. The characteristic Pake pattern of the fully deuterated atactic polypropylene appears twice along the diagonal due to the presence of alkyl deuterons (outer diagonal peaks) and motionally narrowed methyl deuterons (low-intensity, inner diagonal peaks). The broadening of the diagonal into the 2D plane and the absence of elliptical exchange patterns indicates the chain dynamics are isotropic motion by small angular jumps (i.e., rotational diffusion). The lineshape can be analyzed as consisting of subspectra superposed according to a distribution of reorientational angles [13], shown as the inset in Fig. 9a. Shown here for only one mixing time ($t_{mix} = 300$ ms), this spectrum is basically a "snapshot" of the dynamic state of the system and alone provides insufficient information about the motional mechanism. During this mixing time, for example, many different types of motional events may have had time to occur. In fact, this particular spectrum may be simulated either by small-angle reorientations only, *or* by an appropriate combination of small- and large-angle reorientations. It is only with a series of 2D spectra measured for different mixing times that the motional mechanism may be clearly defined [90]. All spectra of such a series must be simulated without changing the respective correlation times of whatever underlying motional processes are modeled. For a fixed temperature, changing an experimental mixing time should obviously not affect correlation times of molecular dynamics. This is an important consideration for all 2D exchange spectra in which the pattern does not cleanly reveal the presence of only one or two discrete processes. The series

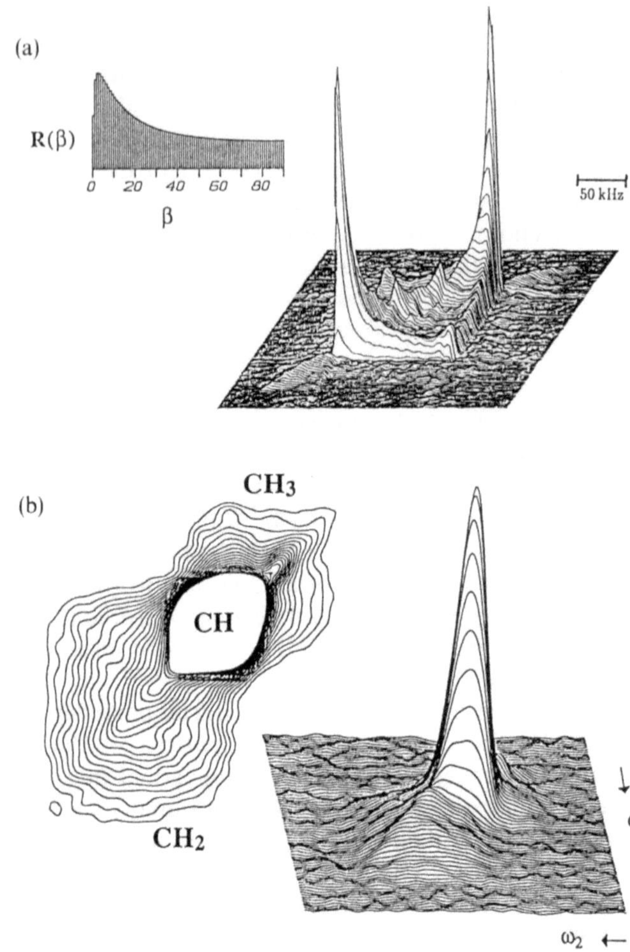


**Fig. 9a, b.** 2D molecular-reorientation exchange spectra for atactic polypropylene (aPP) just above the glass transition ($T_g + 11$ K): **a** $^2H$ spectrum, $T = 264$ K ($T_g$ for this aPP is 253 K), $t_{mix} = 300$ ms, *inset* is the distribution of reorientational angles obtained from the best-fit simulated spectrum [89]; **b** $^{13}C$ spectrum, $T = 250$ K ($T_g$ for this aPP is 239 K), $t_{mix} = 500$ ms [91]. For both spectra, their characteristic forms and lack of elliptical exchange patterns are indicative of diffusive chain reorientations

of $^2H$ 2D exchange spectra of atactic polypropylene (one of which is shown in Fig. 9a) could be simulated with small-angle reorientations only, and not with any combination model [89].

This type of dynamical behavior can also be detected by $^{13}C$ 2D exchange NMR as shown in Fig. 9b [91]. Measured on a static sample, the spectrum diagonal with no mixing time is the overlapping anisotropic chemical-shift lineshapes of the CH, $CH_2$, and $CH_3$ groups. As shown in this spectrum, the diagonal is broadened into the 2D plane, indicative of the isotropic rotational diffusion

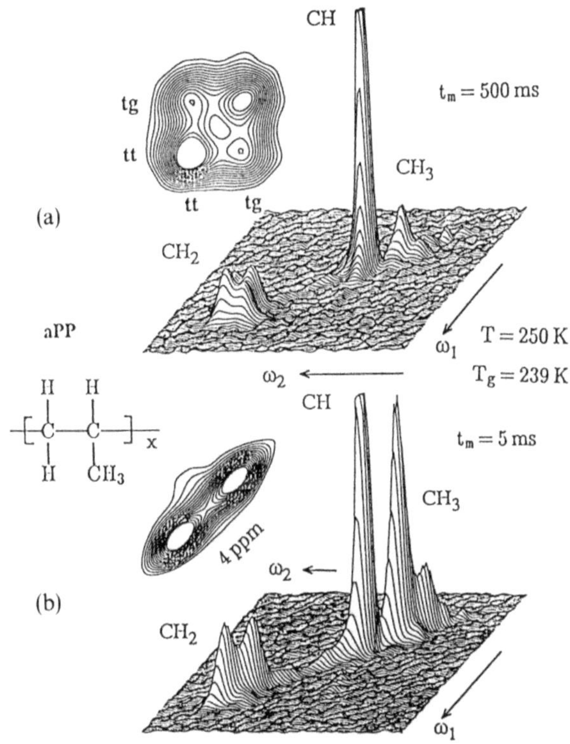


**Fig. 10a, b.** $^{13}C$ 2D MAS exchange spectra for atactic polypropylene just above the glass transition ($T_g + 11$ K): **a** $t_{mix} = 500$ ms, **b** $t_{mix} = 5$ ms. Contour plots of the $CH_2$ chemical-shift regions are shown as *insets* to clearly display the conformational exchange, *trans/trans↔trans/gauche*, for the longer mixing time. The CH peaks are truncated, and the $CH_3$ peaks are diminished due to $T_1$ relaxation [93]

model described for the $^2H$ data. This experiment reveals off-diagonal exchange intensity for changes in the anisotropic chemical shifts caused by changes in the orientations of these CSA tensors with respect to the external magnetic field. At the same time, off-diagonal exchange intensity will also be produced by changes in the isotropic chemical shifts which normally accompany chemical bonding changes, and therefore is traditionally known as chemical exchange [92]. To check for this, the 2D exchange experiment may be performed under conditions of magic-angle spinning, in which the anisotropy of the chemical shift is spun out to reveal the isotropic lineshapes. Figure 10 displays $^{13}C$ 2D MAS exchange spectra for atactic polypropylene [93] at $T_g + 11$ K for mixing times of 500 ms (Fig. 10a) and 5 ms (Fig. 10b).

For the short mixing time, the diagonal is analogous to the 1D MAS spectrum and thus shows the isotropic chemical shifts. Instead of three separate peaks for the three chemically different carbons of polypropylene, splittings and shoulders of peaks are observed which are due to the different conformations

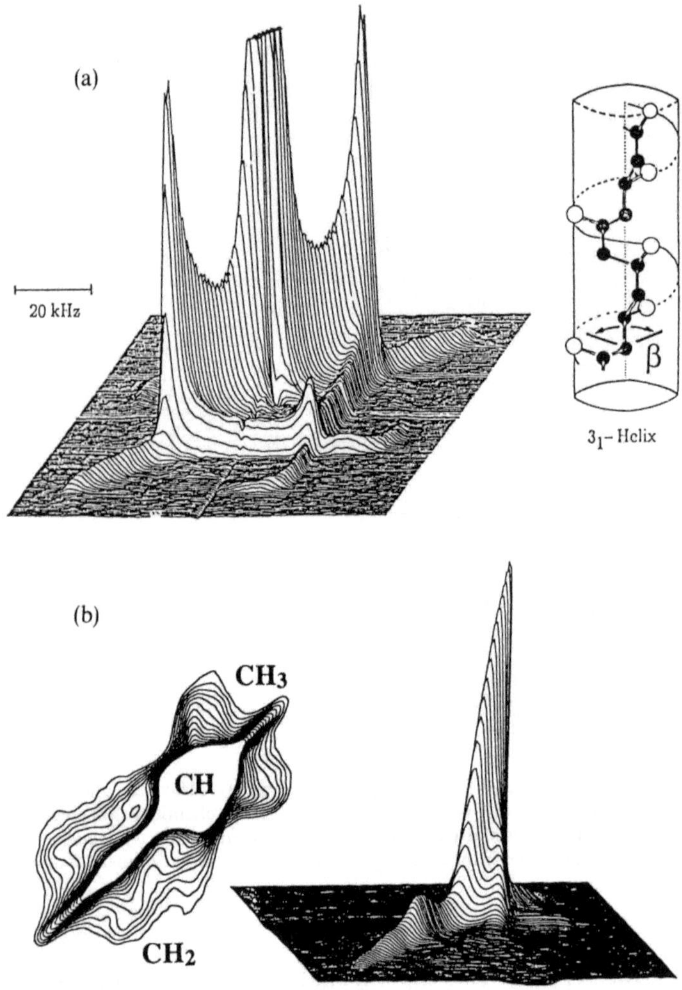


**Fig. 11a, b.** 2D molecular-reorientation exchange spectra for isotactic polypropylene (iPP): **a** $^2$H spectrum, $T = 387$ K, $t_{mix} = 150$ ms, *inset* is the $3_1$ helical structure of iPP, truncated peak is due to methyls of amorphous region [89]; **b** $^{13}$C spectrum, $T = 340$ K, $t_{mix} = 1$ s [88]. For both spectra, the characteristic elliptical ridges are indicative of reorientational angles of 113° that correspond to 120° helical jumps within the crystalline regions

in the solid state. Rotational isomerism is well-known among chain molecules [94] and is directly observed in such solid-state NMR spectra [95]. Especially prominent for the methylene and methyl carbon peaks, the peaks corresponding to carbons with *gauche* carbons in $\gamma$ positions are chemically shifted upfield (see Fig. 10a). On increasing the mixing time, conformational exchange between *trans-trans* and *trans-gauche* sequences are directly observed at this temperature as crosspeaks in the methylene region. At higher temperatures, this conformational exchange becomes so fast that the peaks due to the different sequences

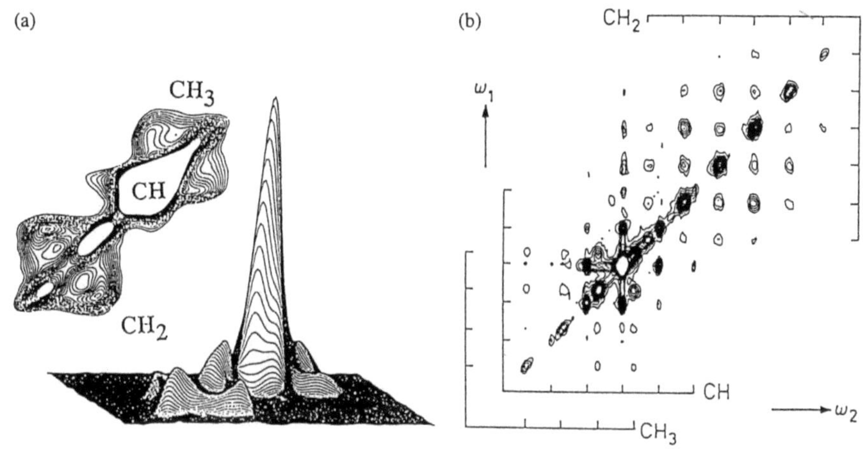


**Fig. 12a, b.** $^{13}C$ 2D molecular-reorientation exchange spectra of isotactic polypropylene: **a** static oriented sample, $T = 365$ K, $t_{mix} = 1$ s [13]; **b** isotropic sample under MAS, $\omega_R/2\pi = 500$ Hz, $T = 340$ K, $t_{mix} = 1$ s [88]. For the static oriented sample, the elliptical ridges indicative of a well-defined jump process are clear; for the isotropic sample under MAS, the sideband intensities must be analyzed to extract the dynamic information

coalesce into one peak along the diagonal [92]. At temperatures below $T_g$, the conformational transitions no longer take place. In fact, on the timescale of ms to seconds, no exchange of any type is observed for this material below $T_g$.

The presence of conformational transitions just above the glass transition indicates that a dynamical description of atactic polypropylene should contain some large-angle reorientations: A *trans* to *gauche* conformational change involves rotations of at least 60° about a chain-backbone bond. On top of this should be added the diffusive-type reorientations by small angles that were clearly detected in the 2D static exchange spectra of Fig. 9. Distributions of correlation times for both of these motional modes provide for quite a complex dynamic behavior in this disordered solid. In contrast, the ordered nature of crystalline arrays strictly limits the types of motions that can take place. Due to bonding and symmetry arrangements, the energetically allowed movements of segments within such an environment are much less than those allowed in a disordered amorphous region. Within the crystalline lamellae of semicrystalline polymers, one such motion proposed for some time is the helical jump [96, 97]. This involves the rotation of the chain about the helix axis by $360° \cdot n/m$ where $m$ is the number of repeat units per $n$ turns of the helix, accompanied by translation of the chain by one monomer unit. Thus, for a single monomer unit, the relative position to the surrounding polymer chains within the crystallite is the same before and after the jump. This type of motion had been proposed for the dynamic mechanical $\alpha$ relaxation of isotactic polypropylene (cf. Fig. 8), although it had never been really proven [97]. Figure 11 shows the 2D $^{13}C$ and $^2H$ exchange NMR spectra of isotactic polypropylene (iPP) that provided the proof that such a motion was indeed possible and does in fact occur [88, 89].

The 2D $^2$H spectrum of iPP selectively deuterated in the methyl groups (Fig. 11a) was taken at 387 K [89]. The cut-off intense peak in the center of the diagonal is due to the motionally narrowed amorphous region. The $3_1$ helical structure of this polymer is shown as an inset. The exchange is revealed as a pair of ellipses with their long axes perpendicular to the diagonal thus indicating a well-defined jump motion. From these characteristic elliptical ridges, the reorientational angle is directly calculated from the 2D plot to be 113° for a well-defined jump process. Taking the chain bonding geometry into consideration, this directly corresponds to the expected 120° helical jump of this $3_1$ helix (360° · 1/3) around the helix axis.

This discrete jump process was first observed in a $^{13}$C 2D exchange spectrum of a static sample of iPP, shown in Fig. 11b [88]. Like the 2D $^{13}$C static spectrum of aPP (Fig. 9b), this spectrum is made up of the three overlapping CSA lineshapes of the three carbons of polypropylene. Although not as clear as in the $^2$H spectrum of iPP, the exchange is suggestive of well-defined large-angle reorientations, and not indicative of some diffusive process defined by a distribution of correlation times (compare with the $^{13}$C static spectrum of aPP in Fig. 9b). The exchange patterns can be made more pronounced by doing the experiment on an oriented sample. This is shown in Fig. 12a for an oriented sample of isotactic polypropylene [13]. Compared to the same spectrum of the unoriented sample (Fig. 11b), the elliptical patterns are clearly detected for all three carbons. This effect occurs due to the concentration of CSA tensor orientations into a smaller surface area, as opposed to the completely random orientations extant in a powder [88, 98]. Thus, depending on the orientation of the sample (and therefore CSA tensors) with respect to the magnetic field, a much higher exchange intensity may result. For example, in a sample oriented such that the CSA orientations are contained within a distribution of 10° width, an exchange intensity increase of about 100 over a powder sample may be produced [99]. However, the exchange patterns for oriented samples depend on the orientation of the samplc with respect to the external magnetic field, and in fact can be quite different. Thus in order to reap the full benefits, 2D exchange spectra for different orientations of the order axis with respect to the external field should be taken.

Although the separate CSA lineshapes overlap for polypropylene, the exchange patterns are still clearly distinguishable for each of the three carbons. However, for more complex materials containing severely overlapping CSA lineshapes, elucidation of exchange patterns can be quite difficult. One method to deconvolute such spectra is with magic-angle spinning. At sufficiently slow spinning speeds, sidebands are retained and therefore also the orientation information useful for these studies of molecular reorientations. However, instead of a direct relationship between the NMR frequency (i.e, the lineshape) and the molecular-segment orientation, this connection is now modulated by the MAS spinning speed [100]. The orientation of all segments varies with the position of the rotor (i.e., phase) in the magnetic field. Thus, to prevent exchange peaks due simply to the rotation of the sample, the position of the rotor must be in the same place at the beginning of the detection period as it was at the end of the evolution period. This is accomplished by synchronizing the pulse sequence with the rotor, and defining the mixing time as an integral number of rotor periods [100]. An example of this experiment is shown in Fig. 12b for isotactic polypropylene spun at 500 Hz [88]. The exchange in this spectrum appears as off-diagonal peaks connecting the sidebands within each respective CSA sideband envelope. The outline of the exchange net recalls the static 2D exchange spectrum of Figure 11b. The improved spectral resolution produced by MAS is counterbalanced by the decrease in reorientational-angle resolution as compared with the static 2D exchange experiment. The highest reorientational-angle resolution is obtained with static lineshapes ($\pm 1°$); this decreases as the number of sidebands decreases, that is, as the spinning speed increases [13].

Veeman and coworkers first developed the MAS 2D exchange experiment, and were the first to demonstrate the existence of the helical jump by examining polyoxymethylene (POM) [101, 102]. This experiment was later improved upon by recording absorptive lines, and applied to reveal the helical jump for isotactic polypropylene [88]. By varying the mixing time in the 2D exchange experiments, correlation times may be estimated (or at least some characteristic time constant of the exchange) by plotting the ratio of exchange intensity to total intensity as a function of mixing time [4]. By varying the temperature in the experiment, activation energies may be obtained from a plot of the correlation times as a function of temperature. It is the similarity of the NMR-measured correlation times and activation energies with those determined by dynamic mechanical and dielectric spectroscopy that provides the final proof that the helical jump is the underlying motion connected with the $\alpha$ relaxation. One or more of these 2D exchange techniques have so far been applied to the following semicrystalline polymers exhibiting $\alpha$ relaxations: polyoxymethylene [102], isotactic polypropylene [88, 89] polyoxyethylene [103], and polyethylene [104]. In all cases, helical jump motions have been shown to occur at temperatures between the glass transition and melting transitions. Molecular model calculations indicate the mechanism involves a "defect" diffusion of the helix rotation through the crystal lattice, as opposed to a rigid-rod rotation of the crystal stem (although this latter process becomes more important as crystal thickness decreases) [105, 106]. These studies are contributing to the emergence of a molecular-level unified description of polymer dynamics.

## 4.3 Dynamics of Other Polymer Systems

### 4.3.1 Chain Diffusion in Polyethylene

The case of polyethylene requires a brief description. Consisting of a $2_1$ helix, the so-called helical jump within the crystalline regions of this polymer consists of a 180° flip of the chain axis accompanied by a translation of one chemical repeat unit [107]. Because the principal axes of the $CH_2$ chemical-shift tensor are aligned parallel and perpendicular to the chain directions [108], rotation by 180° places segments in equivalent positions as far as the static magnetic field is concerned. Thus, the helical jump in polyethylene is not directly observable by 2D exchange NMR. However, it is observable indirectly by the effect produced by the translation step of the helical jump process: that is, chain diffusion between crystalline and amorphous regions [104]. This was detected in a MAS $^{13}C$ 2D exchange experiment as exchange between the isotropic chemical-shift peaks of the amorphous- and crystalline-phase carbons, separated along the chemical-shift axis due to the so-called *γ-gauche* effect [109]. As chains leave one phase and enter the other, the accompanying conformational change (e.g., *trans* to *gauche* upon crystal to amorphous movement) [110] produces the observed exchange peaks. The diffusion was found to be significantly long-range (rms displacements of ~ 10 nm within 100 s at 360 K), and possesses an activation energy of 105 kJ/mol thereby connecting it with the dynamic mechanical α relaxation.

### 4.3.2 Sub-$T_g$ Motions in Poly(methyl methacrylate)

Combinations of 2D exchange techniqes are sometimes needed to reveal the underlying molecular process(es) responsible for mechanically or dielectrically observed relaxations. A particularly interesting example is the β relaxation of poly(methyl methacrylate). Being a completely amorphous polymer, its β relaxation has traditionally been connected with motions of the side group occurring below its glass transition temperature of 105 °C [60]. Recently, the nature of this motion was investigated with $^{13}C$ 2D exchange NMR [111]. Focusing on the carboxyl carbon, evidence of large-angle reorientations was found on a sample at natural abundance. Isotopic enrichment of this carbon to 20% clearly reveals the exchange as shown in Fig. 13a. The geometry of the side group suggests 180° jumps for the large-angle reorientation around the bond connecting the carboxyl carbon to the backbone, but the characteristic elliptical patterns are absent. After confirming the CSA tensor orientation in the molecular frame (shown in Fig. 13d, $x_c \perp$ OCO plane, $y_c$ and $z_c$ are in the OCO plane) [112], it became clear that the observed exchange results from a more complex dynamic. Figure 13b displays the $^2H$ exchange spectrum of PMMA deuterated in the ester-methyl group. Due to rapid rotation around the O—$CH_3$ bond, the electric-field-gradient tensor is aligned with its unique axis along this bond [74]; accordingly, the $^2H$ spectrum is insensitive to 180° flips around this bond alone. The observed exchange in the $^2H$ spectrum reveals the presence of a rocking motion within the OCO plane that is necessarily accompanied by some localized main-chain mobility. When this motion is included with 180° jumps, the $^{13}C$ 2D exchange spectrum may be simulated, as shown in Fig. 13c. Significant is the fact that the side group is asymmetric: Thus, 180° jumps between sites that are energetically inequivalent provide a source for mechanical relaxation. It is important to note that only 50% of the side groups participate in this motion. This fraction correlates closely with the fraction of

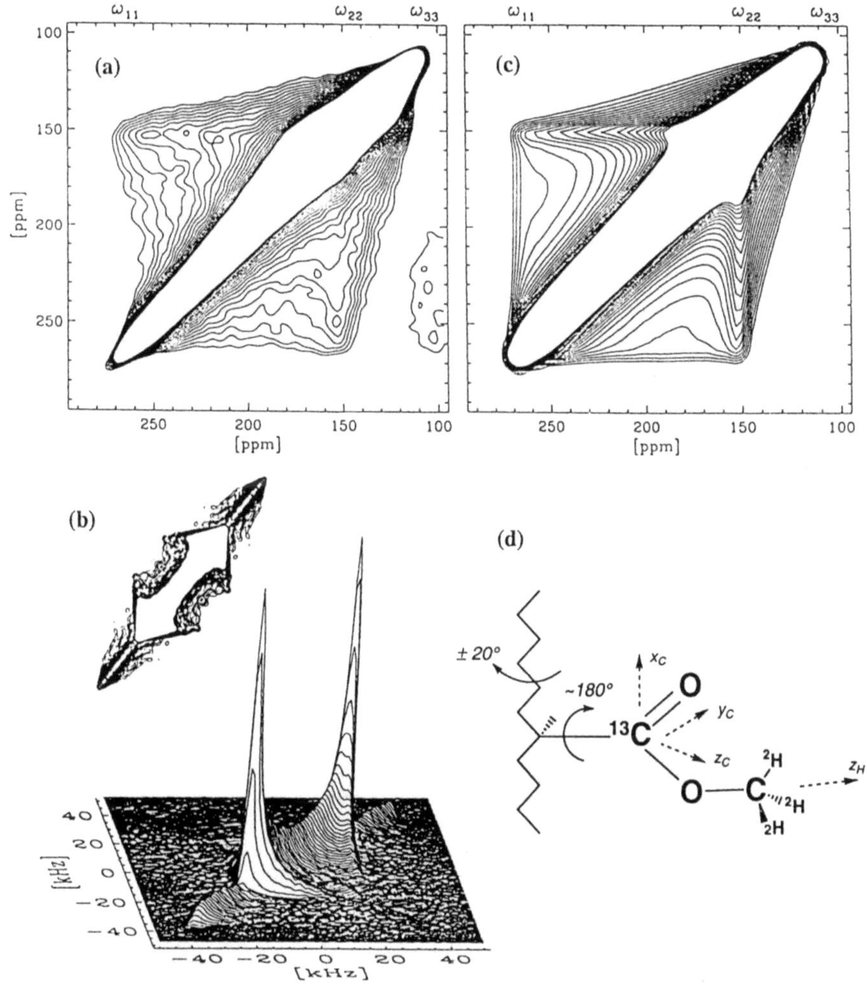


**Fig. 13a–d.** 2D molecular-reorientation exchange spectra revealing the side-group motion of poly(methyl methacrylate): [111] **a** $^{13}C$ spectrum of the isotopically enriched (20%) carboxyl carbon, $T = 333$ K, $t_{mix} = 50$ ms; **b** $^2H$ spectrum of the deuterated ester methyl, $T = 355$ K, $t_{mix} = 500$ ms; **c** best-fit $^{13}C$ 2D exchange spectral simulation for 180°(±10°) flips of the side group, coupled with restricted rocking motions (rms amplitude = ±20°) around the local chain axis (i.e., around $x_c$). **d** Schematic illustrating this motion along with the associated orientations of both the $^{13}C$ chemical-shift tensor [112] and the $^2H$ field-gradient tensor in the molecular structure

syndiotactic triads present in the sample, which give rise to locally extended chain conformations amenable for the rocking motion [111]. Local main-chain torsions that necessarily occur have been postulated before [113, 114] and may turn out to be a general characteristic of $\beta$ relaxations in glassy amorphous polymers with asymmetric side-group motions [115].

### 4.3.3 Jumping Hairy Discotics

Besides the elucidation of molecular dynamics in well-known commercially important polymers, a great deal of work has been done to investigate the variety of motions possible in somewhat more exotic liquid-crystalline polymers [116]. The example presented here is that of phthalocyaninatopoly(siloxane)s (PcPS), which consist of disc-like phthalocyaninato groups attached through their centers to a poly(siloxane) backbone [117]. These structures, shown in Fig. 14, form discotic mesophases and are rendered tractable by radially connecting flexible side chains. The $^2$H 2D exchange NMR spectrum for $t_{mix} = 50$ ms is shown in Fig. 14 and may be simulated by a four-fold jump motion. Thus, the rotational motion of the columns is dictated by the four-fold symmetry of the molecule, and occurs for the individual phthalocyanine discs as opposed to the rotation of the entire column which would be controlled by the six-fold symmetry of the lattice. Although spectra simulated with a model of uniaxial rotational diffusion look rather similar, a number of considerations

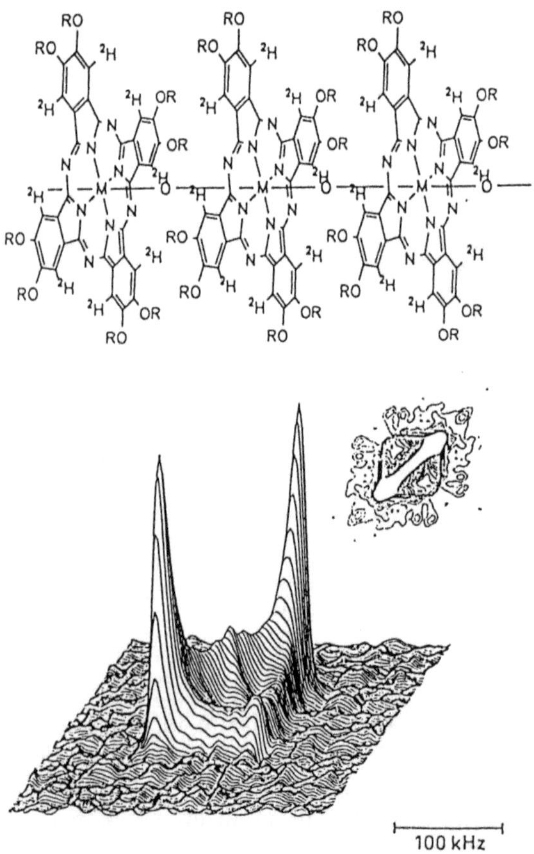


**Fig. 14.** $^2$H 2D molecular-reorientation exchange spectrum of a discotic liquid-crystalline phthalocyaninatopoly(siloxane) (structure shown, $R = C_6H_{13}$): $t_{mix} = 50$ ms and $T = 340$ K. This spectrum may be simulated by a four-fold jump motion around the columnar axis [117]

leads to the conclusion of four-fold jumps. First of all, for a single temperature, spectra assuming uniaxial rotational diffusion cannot be simulated for different mixing times using the same correlation times. This is possible with the four-fold jump model. Secondly, the ratio of exchange intensity to broadened diagonal is about 1:1, which is consistent with a four-fold jump (signal for every second 90° jump appears on the diagonal); for uniaxial diffusion the exchange intensity would be much higher [13]. A spectrum indicative of full exchange via a diffusive process was not found even for the longest possible mixing times.

### *4.3.4 Blend and Network*

The 2D static exchange experiment has been employed to investigate the glass-transition dynamics in a compatible blend of polystyrene and poly(2,6-dimethyl-phenylene oxide) (PXE), in which the PXE was $^{13}$C-labeled [118]. The motion detected was simulated as isotropic rotational diffusion with a bimodal or trimodal distribution of correlation times, depending on the blend concentration. The multimodal correlation time distributions were interpreted with a lattice model in which local concentration fluctuations of each component determined the mobility of each fraction (or mode). The same research group has extended the 2D exchange NMR experiment to the $^{31}$P anisotropic lineshape; diffusive motion was found for a phosphate-ester diluent in a 50:50 blend of PS and PXE [119], as well as for the thiophosphate crosslink in a model network polymer [120]. In the former case, the diluent motion commenced about 20 °C below the caloric glass transition, at the same temperature as the dynamic mechanical loss peak. For the model network, the crosslink motion was connected with the glass transition, described by a stretched exponential correlation function, and exhibited a WLF temperature dependence.

Thus, 2D exchange NMR allows immediate qualitative determinations of the type of slow molecular motions present in solid polymers by simple examination of the 2D maps. More detailed quantitative information such as the fraction of the sample participating in the dynamic process being monitored, or the distribution of reorientational angles defining an isotropic diffusive motion may be extracted by a complete lineshape analysis coupled with simulations [13].

## 5 Order: Exchange by Sample Reorientation

Preferential orientation of polymer chains, either intentional, accidental, or natural, has long been known to significantly affect the bulk physical properties of otherwise isotropic materials. Quantifying the direction and degree of preferred orientation, or texture, is obviously necessary to relate molecular structure to macroscopic properties so that appropriate processing schemes may be devised [121]. In the three-dimensional space in which materials exist, any preferred orientation must be quantified as the probability of finding a specified molecular-segment direction, within a defined tolerance, relative to one of three orthogonal directions assigned to the macroscopic sample. The specified molecular-segment directions will be distributed around average deviations from these three sample directions. Transformations between the molecular frame and the sample-fixed frame are defined by three Euler angles $(\theta, \phi, \psi)$ for coordinate axis rotation: Successive rotations about these angles place the coordinate axes into coincidence [13]. The orientation-distribution function $\rho$ is simply expressed as $\rho(\theta, \phi, \psi)$. The situation is greatly simplified in polymer fibers [122]. The three assigned sample directions are logically chosen as the fiber direction and two perpendicular directions. Obviously, since fibers possess cylindrical symmetry, distinctions between any two directions perpendicular to the fiber direction are meaningless. Thus, orientation may be defined in terms of one angle: It is the distribution of a specified molecular-segment direction (typically the chain axis for polymers) with a deviation of $\theta$ from the fiber axis. The orientation-distribution function simply becomes $\rho(\theta)$. In biaxially oriented and bulk oriented materials, the quantification of orientation is significantly more involved as all three Euler angles must be specified. The situation can become even more difficult to analyze for molecularly heterogeneous materials in which different molecular components or segments can orient separately. Each component or segment must then be defined in terms of its three Euler angles for a full description of the molecular orientation of the sample. For this, techniques must exist for separately measuring the orientation of the different segments.

## *5.1 Measurement of Orientation*

In general, appropriate analytical techniques provide quantifiable measures related to macroscopic sample directions. With knowledge of the geometrical relation of the quantifiable measure to the molecular framework, the connection can be made between the molecular framework and the macroscopic sample directions. All that is needed is some measurement sensitive to the molecular anisotropy. A number of techniques have been developed which provide orientation information [121, 122]. Examples include optical birefringence and polarized fluorescence which give orientation information averaged over the entire sample [122]. Sensitive primarily to crystalline regions, X-ray diffraction has been extensively employed for characterizing the direction and degree of orientation in polymeric materials, most completely with pole-figure analysis [123]. In X-ray diffraction, the angle between the crystallite planes and the molecular framework must be known. This, in turn, requires that something be known about the crystal structure of the material. For polymers with sufficiently symmetrical crystal structures, some reflection can usually be identified along the chain-axis direction. Although some information is provided about ordering within amorphous phases, this technique is most powerful for crystalline arrays in which a regular, periodic structure gives single diffraction peaks, the intensity of which depend on the orientation of the ordered sample with respect to the incident X-ray beam [123]. Spectroscopic measurements have the distinct advantage of being able, in favorable cases, to provide orientation information on both amorphous and crystalline regions. Examples are infrared (IR) dichroism [124], laser-Raman spectroscopy [125], and solid-state NMR spectroscopy [14]. Due to the molecular specificity of these techniques, orientation information on individual molecular moieties may be obtained for the same material. This has proved extremely valuable, for example, in segmented or articulated polymer systems in which different parts of the same molecule orient independently [6].

All of the techniques mentioned above have been employed for orientation measurements with varying degrees of success depending on the nature of the material under investigation. With two exceptions, the methods provide the orientation information in terms of one or more "orientation averages" [122]. The orientation averages are reported as order parameters, which are the expectation values of the corresponding terms of an expanded orientation distribution. Uniaxial distributions are normally expanded in terms of Legendre polynomials because they are orthogonal, which facilitates often encountered mathematical operations for interrelating distributions or rotating reference axes (e.g., in transforming from the molecular-fixed coordinate frame to the sample-fixed coordinate frame) [13]. These values can only be related to moments of the desired orientation distribution. Adopted from mechanics, the term "moment" was first used to indicate importance. The first moment is related to the mean, the second moment to the breadth, the third moment to the symmetry or skewness, and the fourth moment to the "peakedness" of a normalized distribution [126]. In many cases, knowledge of the moments of the orientation distribution is enough information to adequately describe the order of a system [127]. Because a Legendre expansion does not converge for highly ordered systems (full-width-at-half-maximum $< 10°$), narrow orientation distributions are often described simply by the root-mean-square width of an assumed Gaussian [13].

When these approximations are insufficient, X-ray and solid-state NMR methods may be employed to yield the complete orientation distribution (that is, as long as the NMR lineshape being investigated is due to a single-spin intramolecular coupling such as CSA or quadrupolar couplings; dipolar couplings, measured by $^1$H wideline NMR for example, are multispin interactions from which only the moments of the distribution can be extracted [128]). The differences in the two cases, those methods that provide moments and those methods that provide the complete distribution, may be generalized in terms of the manner in which angular information (orientation) is encoded in the quantifiable measure. To provide the complete distribution, the quantifiable measure must be a direct image of the orientation distribution. That is, each intensity point of the quantifiable measure must correspond to a single angular point in the distribution, or at least to a small number of angular points related to each other in a known manner. Furthermore, each intensity point must represent the weight fraction of the molecular segment oriented at that respective angle. This is the case in X-ray diffraction, and also in certain solid-state NMR methods in which the angular information is directly contained across the NMR frequency spectrum. For noncentrosymmetric structures, X-ray diffraction data are true mass-fraction-versus-angle pictorializations (i.e., *the* orientation distribution). X-ray patterns for the more common centrosymmetric structures, and solid-state NMR data in general, are symmetrized. In NMR, this is due to the magnetic equivalence of nuclei whose CST axes are directed 180° apart. Thus if the intensity is convoluted by nonrelated or redundant information, or more relevantly, if the intensity at a single angular point contains contributions from all other directions, then the quantifiable measure does not represent the orientation distribution.

To illustrate, a brief comparison will be outlined for the measurement of the chain-axis distribution along a fiber (only one angle needs to be specified) with infrared dichroism and solid-state NMR of a CSA lineshape (spin 1/2). With IR dichroism, only the second moment of the orientation distribution can be obtained [124]. There is no frequency encoding of the angular information; all information is contained in the intensity of a chosen spectral band that represents a specific molecular segment. The intensity of this band varies with the direction of polarization of incident electric field. It is greatest when the incident electric field is parallel to the transition dipole moment (the product of these two vector quantities is effectively integrated over all directions to give the intensity). Thus, the intensity of a band contains contributions from all molecular segments at all orientations. On the other hand, with solid-state NMR of an anisotropic chemical-shift lineshape, the angular information is encoded in the NMR frequencies [3]. At each frequency (i.e., angle), the intensity is related to the weight fraction of molecular segments at that particular frequency (or angle). After taking the aforementioned symmetry into consideration, the orientation distribution is directly reflected in the lineshape.

## *5.2 1D NMR*

The above is true in solid-state NMR only for nonoverlapping static lineshapes of axially symmetric ($\eta = 0$) spin-1/2 nuclei. When there exists overlap due either to other chemical species (e.g., as in most $^{13}C$ spectra of static samples), or to the other "transition" of a spin-1 nucleus (e.g., as in all $^{2}H$ spectra), direct reconstruction of an orientation-distribution function is hindered. In addition, the nonlinear relationship between NMR frequency and orientation angle (see Sect. 2) means the angular resolution is extremely low at the singularities of a wideline spectrum (i.e., the edges or maxima), and thereby makes it difficult to extract a unique distribution from a single experimental spectrum [13]. One-dimensional $^{2}H$ wideline NMR has been extensively employed for devolving molecular-level orientational information in fiber-forming [129–132] and liquid-crystalline [133, 134] polymers. In practice, a set of so-called rotation patterns is measured which is simply a series of 1D $^2H$ spectra taken at different angles between the sample and magnetic field. Matching the series of spectra with simulations allows analysis of the data. However, unique orientational distribution functions are produced only in the case of simple uniaxial distributions: Only one Euler angle can be specified. Despite the overlap and low-angular-resolution-at-the-singularities problems, the 1D $^2H$ NMR technique has been particularly successful for the following reasons: the extraction of orientation information from different parts of the same molecular structure, and the ease of data analysis due to $\eta = 0$ (in most cases) coupled with the coincidence of the interaction-tensor principal axis with the C—$^2H$ bond. Other lineshapes analyzed for orientational information are the $^{19}F$ CSA in poly(tetrafluoroethylene) [135, 136], and the $^{13}C$ CSA in a polycarbonate [137] and a polyimide [138].

## *5.3 2D NMR*

Both primary problems associated with the 1D lineshape studies of orientation are greatly reduced in a 2D exchange experiment [139]. The sample is rotated during the mixing time of the experiment (cf. Fig. 2) so that the NMR-frequency-encoded angular information is correlated, and the data are presented in two dimensions. This has been demonstrated for deuterium [140] as well as $^{13}C$ nuclei [141] in an experiment most recently called DECODER, or Direction Exchange with Correlation for Orientation-Distribution Evaluation and Reconstruction. The experiment is technically similar to one devised for determining or disentangling the principal values of chemical-shift tensors; principal values of CST's are determined with greater accuracy due to the expanded frequency presentation in two dimensions which reduces overlap [142]. It is exactly analogous to the 2D exchange techniques described previously. In this case, the exchange signals are created by molecular-segment orientation changes due to movement of the entire sample within the external magnetic field during the mixing time. In the resulting 2D spectra, the projections onto the two frequency axes simply correspond to the 1D spectra for those respective orientations. The 2D correlation of two orientation-dependent frequencies allows the determination of two angles specifying segmental orientations. Exchange due to spin diffusion or molecular mobility must be either nonexistent, insignificant, or corrected. Relevant cross-sections through the orientation distribution have been reconstructed directly from the spectral intensity [13,141]. Thus, the DECODER experiment is well suited to samples that exhibit more than simple fiber symmetry. All three Euler angles needed for a full description of molecular orientation could be specified in a 3D DECODER experiment with two mixing times in which the sample is rotated twice and all three positions are correlated via the orientation-dependent NMR frequencies [139, 141]. The need for a three-dimensional measurement to fully elucidate molecular orientation profiles has been similarly recognized with IR spectroscopy [143].

### 5.3.1 *Comparison to X-ray Diffraction*

Rotating the sample and correlating measurements to define orientation is also the routine followed in X-ray pole-figure analysis for measuring orientation [123]. The similarities between DECODER NMR and X-ray scattering go beyond simple procedural considerations and have been carefully outlined [140]. The advantages of NMR are its molecular specificity and the unnecessary existence of crystalline arrays; thus, molecular orientation in amorphous materials may be characterized. The similarities are even apparent in the raw data produced, at least for certain experimental arrangements. That is, in both techniques, the existence of texture is reflected by decreased spreading of the spectral intensity. For example, consider a flat-film X-ray photograph of an isotropic polymer exposed to a pinhole beam. The pattern will consist of concentric circles of constant circumferential intensity (called Debye rings), with radii corresponding to the *d*-spacings of the major reflections [144]. For a circle of fixed radius (i.e., a fixed *d*-spacing, and therefore a specific crystalline reflection), the intensity *around* the 360° circumference represents the variation in a single Eulerian angle while holding two others constant. This is called an azimuthal scan [145]. For the isotropic sample, a plot of intensity versus azimuthal angle will be a straight horizontal line. Replacing the isotropic sample with a slightly oriented sample of the same polymer produces a pattern characterized by circles of nonuniform circumferential intensity. For highly oriented fibers, the circles are replaced by arcs with lengths dependent on the degree of orientation of the polymer chains in the fiber; in the limit of perfect orientation, the arcs will simply contract to points on the radii of the former circles. For the oriented fiber, a plot of intensity versus azimuthal angle will consist of two peaks separated by 180°, the widths of which directly reflect the arc lengths and therefore the degree of polymer orientation. It is a direct measure of the orientation distribution of the normals to the crystallite planes represented by that reflection [123]. Geometrically relating these normals to the molecular structure provides the molecular orientation information.

Now consider an isotropic polymer with a single resonance and $\eta \approx 0$ rotated through 45° during the mixing time of a DECODER experiment so that the unique axis of the principal axis system is rotated in the plane of the external magnetic field. The resulting pattern will consist of a circle with two tangential ridges parallel to the frequency axes, exactly as displayed in Fig. 3. Direct reconstruction of the orientational distribution can be accomplished by making the frequency-to-angle transformation [13]. For the isotropic sample, a $\rho(\theta)$ versus $\theta$ plot (for example) would be a horizontal straight line. The spectral intensity contained in the circular (or elliptical for flip angles other than 45°) ridge corresponds to the field plane, the plane containing the external magnetic field and perpendicular to the sample flip axis. A highly oriented fiber of the same polymer arranged with its order (fiber direction) axis within the field plane would yield a 2D DECODER spectrum in which the circular ridge has been replaced by an arc. The length of the arc depends on the degree of orientation and the location of the center of the arc around the circumference of the circle depends on the average deviation of the PAS unique axis from the sample order axis. If the $z$ axis of the PAS (the unique axis for $\eta = 0$) were parallel to the polymer chain axis, then the transformed reconstructed orientation distribution would be the desired distribution.

### *5.3.2 Poly(ethylene terephthalate)*

For technique demonstration, $^{13}C$ DECODER spectra will be presented for both an unoriented sample as well as uniaxially oriented fibers of poly(ethylene terephthalate) [141]. Figure 15 shows both the contour and stacked plots of an isotropic bulk sample of PET rotated through an angle of 40° during the mixing time of a DECODER experiment. The simulated spectra are also included. The elliptical ridges are present for all carbons of the PET structure; the most obvious are the patterns for the axially symmetric $OCH_2$ (furthest upfield, upper right in contour plot) and carboxyl carbons (furthest downfield, lower left in contour plot). In the stacked plots, relative intensities are more clearly depicted; the much more intense signal arises from the intersections of the ridges running parallel to the frequency axes. The overlapping patterns are typical of $^{13}C$ NMR on static samples with more than one different type of carbon; this can be circumvented by using $^2H$-labeled polymers in such studies [140]. To illustrate what happens to the elliptical ridges when an oriented sample of the same polymer is observed with DECODER NMR, Figure 16 displays the $^{13}C$ 2D DECODER spectrum for a bundle of uniaxially oriented fibers of PET [141]. Focusing attention on the $OCH_2$ and carboxyl portions of the spectrum, the experimental and simulated spectra of Fig. 16 should be compared with those of the unoriented sample in Fig. 15. As described earlier for the hypothetical example, the effects of orientation are manifested by contractions of the elliptical ridges into arcs, the length of which are dependent on the degree of orientation. This spectral intensity was transformed from the $OCH_2$ pattern to yield the distribution of PAS unique axes shown in Fig. 16b. Using an angle of $\sim 8°$ between the PAS unique axis and the polymer-chain axis (determined from a separate DECODER experiment), this distribution was transformed to yield the distribution of chain axes in Fig. 16c. The corresponding distributions for the simulated spectrum are presented on the right side of the figure, confirming the validity of both the simulation and the transformations.

The distribution of chain axes shown in Fig. 16c was reported to have a full-width-at-half-maximum (fwhm) value of $23 \pm 5°$. This highly oriented component is represented in the $OCH_2$ pattern by the really intense signal arced along the right side of the elliptical contour of Fig. 16a (upper right, and black

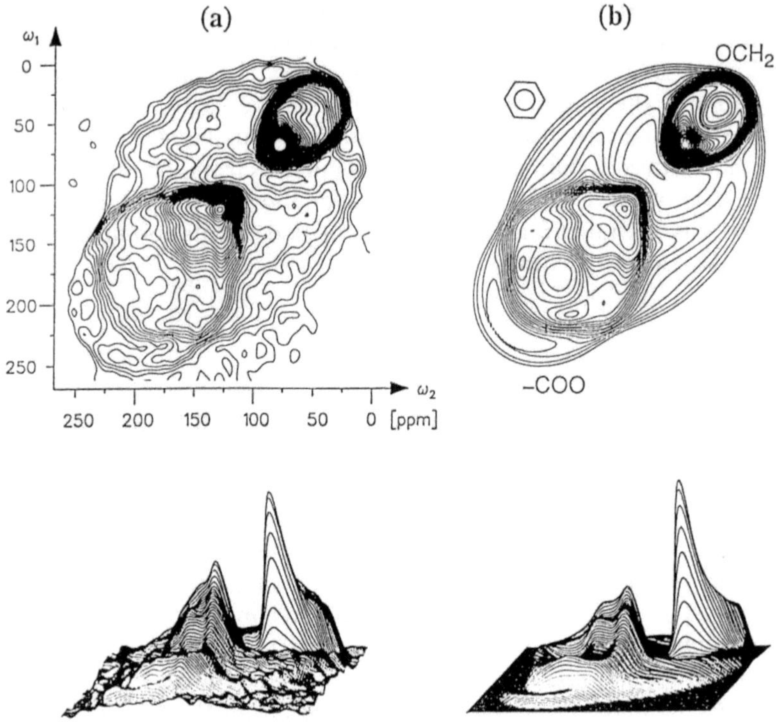


**Fig. 15a, b.** $^{13}$C 2D DECODER spectra of isotropic poly(ethylene terephthalate) flipped through an angle $\beta = 40^\circ$ during the mixing time. Contour plots are above the stacked plots: **a** experimental spectrum, **b** simulation. In the simulated contour plot, spectral areas associated with the different structural components are marked; the $OCH_2$ pattern is in the upper right-hand corner and the carboxyl pattern is in the lower left-hand corner [141].

in contour plot). Also evident in this spectrum is low-intensity signal spread out from the elliptical ridge (gray in the contour plot) indicative of disordered segments of the sample. From the simulation, this fraction of the sample is reported to be distributed with a fwhm of $135 \pm 45^\circ$, which may not be static disorder alone. It appears as background in the reconstructed orientation distributions. It should be noted that the reconstruction is not unique for about 135° to 170°. This is because of signal overlap from different orientations that can occur along the elliptical ridge near a singularity [140]. This region is depicted by dashed lines in the angular distributions of Fig. 16.

Thus, even though the 2D DECODER technique is a significant advancement compared to 1D NMR spectra, the entire distribution is not always accessible by direct reconstruction. However, as exemplified by the distributions shown in Fig. 16, this is not at all a big problem. By judicious choice of flip angle and sample arrangement in the magnet, rather complex distributions may be characterized. The technique is particularly valuable for complex systems containing more than one type of orientation distribution. This has been

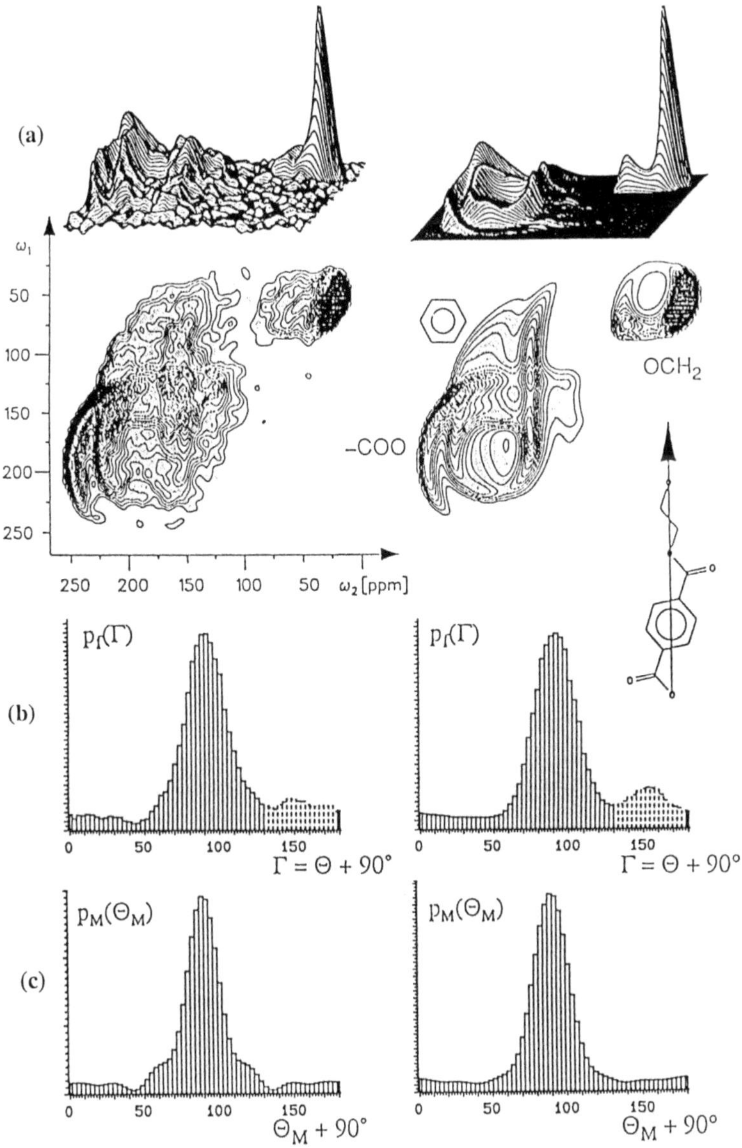


**Fig. 16a.** $^{13}C$ 2D DECODER spectrum of uniaxially oriented poly(ethylene terephthalate) fibers flipped through an angle $\beta = 42^\circ$ during the mixing time. The corresponding simulated spectrum is included on the right. In the simulated contour plot, spectral areas associated with the different structural components are marked; the $OCH_2$ pattern is in the upper right-hand corner and the carboxyl pattern is in the lower left-hand corner. **b** Distribution of PAS unique axes reconstructed from the $OCH_2$ pattern. **c** Distribution of polymer chain axes (fwhm $= 23 \pm 5^\circ$) calculated from the distribution in **b**. The corresponding distributions from the simulation are also included on the right [13, 141]

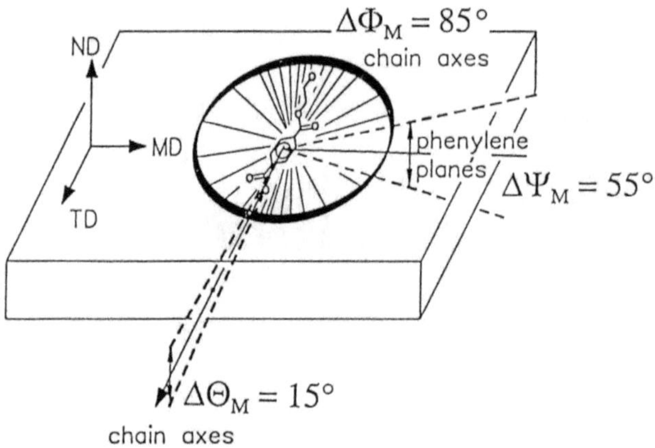


**Fig. 17.** Summary of chain-orientation distributions in biaxially oriented PET films as determined from 2D DECODER spectra. The distributions are represented by their full-width-at-half-maxima: $\Psi_M \Phi_M \Theta_M$ are the Euler angles specifying the molecular-chain-axes orientation in the sample-fixed frame. The sample-fixed frame is defined in reference to the manufacture of the PET films: *MD* is the machine direction, *TD* is the transverse direction, and *ND* is the direction normal to the film plane [141]

convincingly demonstrated by the complete characterization of the molecular orientation distributions in biaxially oriented PET films [141]. Some results are summarized in Fig. 17 in terms of the fwhm for key distributions; the level of detail is a hallmark of NMR spectroscopy. In general, the chains were found to be preferentially oriented parallel to the surface of the film, with the chain axes aligned more in the direction transverse to the machine direction. The specifics are shown in Fig. 17 and described in the figure caption.

### *5.3.3 MAS Methods*

For completeness and perspective, an alternative 2D NMR method for characterizing orientation in polymers will be briefly described that is not an exchange-type experiment. This technique is based on the signal periodicity evolved when spinning a sample with partial orientation [146]. As long as the sample does not exhibit cylindrical symmetry around the rotor axis (e.g., a fiber is not packed in the rotor with the fiber axis parallel to the rotor axis), the NMR spectrum will depend on the phase of the rotor. The alignment of the ordered sample with respect to the magnetic field, and thus the anisotropic resonance lineshape, will be equivalent only at the beginning and at the end of each rotor cycle. Thus, the experiment is quite simply performed by measuring a series of spectra with different initial rotor phases [147]. For each spectrum, every scan must begin at the same rotor phase, which necessitates synchronization of the rotor with the pulse sequence. The total range of $t_1$ is a rotor cycle, which is typically incremented 8 or 16 times to give the corresponding number of

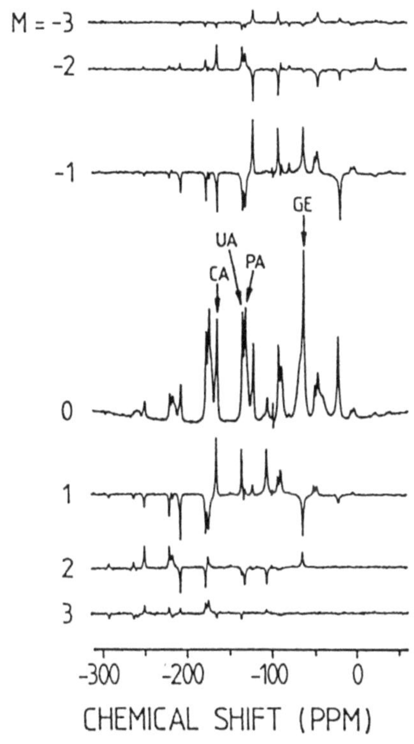


**Fig. 18.** $^{13}C$ 2D MAS spectrum of uniaxially oriented PET fibers. The centerband slice ($M = 0$) is parallel to $\omega_2$ and contains the peak assignments: CA = carbonyl, UA = unprotonated aromatic, PA = protonated aromatic, GE = glycolic ethylene. For each of these peaks, the slices at $\pm M\omega_R$ contain the respective sidebands (in $\omega_1$) which are analyzed to extract the orientational information [147]

slices in $\omega_1$ after the second Fourier transformation "against the rotor phase." Besides rotor synchronization, the experiment requires no specialized hardware to implement (DECODER requires a sample-flipping probehead). Figure 18 shows a 2D data set for oriented PET fiber [147]. The orientational information is contained in $\omega_1$ and is extracted by analyzing the sideband intensities. These sideband intensities can be calculated analytically for perfectly oriented segments. The deviations from perfect orientation are determined using a linear regression method, and expressed in terms of coefficients related to the orientational distribution function. In addition to PET fibers [147, 148], the technique has been applied to stretched samples of amorphous polycarbonate [149] and poly(methyl methacrylate) [112], as well as to oriented fibers of semicrystalline nylon 6 [167], ultrahigh-molecular-weight polyethylene [150], and poly(*p*-phenylene terephthalamide) [151]. For low degrees of order, this experiment may be more appropriate than DECODER since it directly detects deviations from cylindrical symmetry.

With increasing structural, and therefore spectral, complexity of systems under investigation, problems arise due to overlap of the sidebands needed for analyses. This problem has been addressed by the introduction of a 2D experiment that takes advantage of the signal periodicity introduced into the anisotropic part of the chemical shift by spinning the sample [152]. The sidebands are separated along the $\omega_1$ dimension according to their order number. When applied to isotropic samples exhibiting hopelessly inextricable sideband patterns, the experiment sufficiently resolves the spectrum in two dimensions so that the principal components of the chemical-shift tensors can be determined [153]. This technique was quickly combined with the 2D MAS order experiment to produce a rotor-synchronized three-dimensional MAS method for evaluating orientational distribution functions in spectrally complex systems [154]. The 3D technique (named ORDER, for ORientation Distributions with Enhanced Resolution) was originally demonstrated on a smectic liquid-crystalline side-group polymer and most recently on poly(*p*-phenylene terephthalamide) [155]. For the latter polymer, the results from the 3D MAS data were compared with information retrieved from a 2D DECODER experiment: Identical values for the orientation distribution were found. It was noted that although the 3D technique yields inherently better resolved spectra (MAS and three dimensions for spreading out information), the 2D DECODER method is more precise for determining orientation distributions.

## 6 Outlook

The foregoing presentation has concentrated on a selected few experiments which yield valuable information on the bulk material properties of solid polymers. As mentioned in the Introduction, the investigation of polymers with NMR is a huge subject requiring numerous books for adequate coverage [11–14]. The techniques described here were chosen for inclusion because they provide information valuable to polymer and materials scientists, and therefore should promote communication and collaboration between these groups and spectroscopists. It has been the synergism created by such working relationships that has led to most of the results presented in this chapter.

With the increasing availability of instrumentation and expertise, the application of new solid-state NMR techniques will grow. Most of the advanced multidimensional techniques have been performed on relatively simple synthetic polymers [13]. Applications to naturally occurring polymers [156], inorganic polymers [157], and solid-state chemistry [158] have not yet been fully exploited. At the same time, technique development will no doubt continue. Not specifically covered here are related experiments that correlate two of the three named material-property descriptors of solid polymers. For example, 2D and 3D experiments have been devised which correlate order with dynamics [159]. Current method development continues to focus on extending these experiments to more complex materials by the introduction of additional dimensions [13] (e.g., the 3D ORDER experiment) [154] and clever manipulation of nuclear interactions via pulse sequences and specialized probeheads [160]. With the development of such probeheads, the influence of such parameters as tensile stress on molecular mobility reveals important correlations [161]. Furthermore, by spectroscopic solid-state imaging [162], materials properties such as molecular order and dynamics can be measured in a spatially selective manner.

The utilization of quadrupolar nuclei and low-sensitivity nuclei in materials properties studies will increase as usable techniques are improved or become available. Advances in instrumentation are likely to make experiments exceedingly easy. With the advent of very fast MAS, the direct extension of liquids-NMR techniques to the solid state might become more common. As represented best by the three-dimensional structure elucidation of complex proteins with multidimensional COSY and NOESY experiments, this methodology promises to reveal the same type of structural detail in solid materials. Allusions of this appear in studies on concentrated solutions of polymers in which the gain in segmental mobility through localized dissolution precludes the need for MAS to achieve resolution [163]. It was argued that the specific interactions revealed in the concentrated polymer solutions could be correlated to those influencing solid-state properties. Other multidimensional correlation experiments inspired by liquids-type spectra have been reported in which the elimination of broadening interactions is a primary component [164, 165]. Thus, by developing techniques and hardware, the synthesis and manipulation of advanced materials through molecular-scale engineering [166] will be analytically supported.

*Acknowledgments.* This review is based on the work of the NMR group at the Max-Planck-Institut in Mainz. The contributions, imagination, hard work, and perseverance of coworkers are gratefully acknowledged: Bernhard Blümich, Christine Boeffel, Lothar Brombacher, W. Z. Cai, Brad Chmelka, Joachim Clauss, Norbert Egger, Stefan Féaux de Lacroix, Alfred Hagemeyer, Michael Hansen, Gerard Harbison, Manfred Hehn, Sigmar Kaufmann, Andrzej Kulik, Johannes Leisen, Volker Macho, Angela Ohlemacher, Ute Pawelzik, Dieter Schaefer, Claudia Schmidt, Jeremy Titman, Der-Lii Tzou, Stefan Wefing, Michael Werth, Manfred Wilhelm, Yuliang Yang, and Karen Zemke. Special thanks are extended to Klaus Schmidt-Rohr for his critical reading of the manuscript and his steadfast advice on all matters scientific. Comments on some sections from Prashant Desai of Georgia Tech are duly appreciated. Financial support (HWB) was provided through a NSF-NATO postdoctoral fellowship in science and engineering.

## 7 References

1. Kroschwitz JI (1990) Concise encyclopedia of polymer science and engineering. Wiley, New York
2. Brandrup J, Immergut EH (1989) Polymer handbook. Wiley, New York
3. Mehring M (1983) Principles of high resolution NMR in solids. Springer, Berlin Heidelberg New York

4. Ernst RR, Bodenhausen G, Wokaun A (1987) Principles of magnetic resonance in one and two dimensions. Clarendon Press, Oxford
5. (a) Frydman L, Chingas GC, Lee YK, Grandinetti PJ, Eastman MA, Barrall GA, Pines A (1992) Israel J Chem 32: 161; (b) Bax A, Szeverenyi NM, Maciel GE (1983) J Magn Reson 52: 147; (c) Gan Z (1992) J Am Chem Soc 114: 8307
6. Blümich B, Spiess HW (1988) Angew Chem Int Ed Engl 27: 1655
7. Spiess HW (1991) Chem Rev 91: 1321
8. Spiess HW (1991) Annu Rev Mater Sci 21: 131
9. Chmelka BF, Schmidt-Rohr K, Spiess HW (1994) In: Tycko R (ed) NMR probes of molecular dynamics. Kluwer, Minneapolis
10. Spiess HW, Schmidt-Rohr K (1994) In: Richert R, Blumen A (eds) Disorder effects on relaxational processes. Springer, Berlin Heidelberg New York
11. Komoroski RA (ed) (1986) High resolution NMR spectroscopy of synthetic polymers in bulk. VCH, New York
12 Koenig JL (1992) Spectroscopy of polymers. American Chemical Society, Washington
13 Schmidt-Rohr K, Spiess HW (1994) Multidimensional solid-state NMR and polymers. London, Academic Press
14. McBrierty VJ, Packer KJ (1993) Nuclear magnetic resonance in solid polymers. Cambridge University Press, Cambridge
15. Abragam A (1961) The principles of nuclear magnetism. Oxford University Press, Oxford
16. Pake GE (1948) J Chem Phys 16: 327
17. Schaefer J, McKay R, Stejskal E, Dixon WT (1983) J Magn Reson 52: 123
18. Schaefer J, Sefcik MD, Stejskal EO, McKay RA, Dixon WT, Cais RE (1984) Macromolecules 17: 1107
19. Schaefer J, Stejskal EO, Perchak D, Skolnick J, Yaris R (1985) Macromolecules 18: 368
20. Schaefer J, Stejskal E, McKay R, Dixon WT, (1984) Macromolecules 17: 1479
21. Schaefer J, Garbow J, Stejskal E, Lefelar JA, (1987) Macromolecules 20: 1271
22. Linder M, Höhener RR, Ernst RR (1980) J Chem Phys 73: 4959
23. Nakai T, Ashida J, Terao T (1989) Magn Reson Chem 27: 666
24. Schmidt-Rohr K, Wilhelm M, Johansson A, Spiess HW (1993) Magn Reson Chem 31: 352
25. Veeman WS (1984) In: Progress in NMR Spectroscopy 16: pp 193–235
26. Schaefer J, Stejskal EO, Buchdahl R (1975) Macromolecules 8: 291
27. Schaefer J, Stejskal EO (1976) J Am Chem Soc 98: 1031
28. Jeener J, Meier BH, Bachmann P, Ernst RR (1979) J Chem Phys 71: 4546
29. Schmidt C, Wefing S, Blümich B, Spiess HW (1986) Chem Phys Lett 130: 84
30. Hagemeyer A, Brombacher L, Schmidt-Rohr K, Spiess HW (1990) Chem Phys Lett 167: 583
31. Wefing S, Spiess HW (1988) J Chem Phys 89: 1219
32. Dagani R (1992) Chemical & Engineering News 70(47): 18
33. Cheung TTP, Gerstein BC (1981) J Appl Phys 52: 5517
34. Cheung TTP (1981) Phys Rev B 23: 1404
35. Goldman, M, Shen L (1966) Phys Rev 144: 321
36. Assink RA (1978) Macromolecules 11: 1233
37. Caravatti P, Neuenschwander P, Ernst RR (1986) Macromolecules 19: 1889
38. Schmidt-Rohr K, Clauss J, Blümich B, Spiess HW (1990) Magn Reson Chem 28: S3
39. Edzes HT, Bernards JPC (1984) J Am Chem Soc 106: 1515
40. Colombo MG, Meier BH, Ernst RR (1988) Chem Phys Lett 146: 189
41. Robyr P, Meier BH, Ernst RR (1989) Chem Phys Lett 162: 417
42. Henrichs PM, Linder MJ (1984) Magn Reson 58: 458
43. Tycko R, Dabbagh G (1992) Israel J Chem 32: 179
44. Schmidt C, Blümich B, Spiess HW (1988) J Magn Reson 79: 269
45. (a) Spiess HW (1993) Ber Bunsenges Phys Chem 97: 1294; (b) Féaux de Lacroix S (1992) Ph.D. Thesis, Mainz-Johannis Gutenberg Universität
46. Caravatti P, Neuenschwander P, Ernst RR (1985) Macromolecules 18: 119
47. Belfiore LA, Graham H, Veda E, Wang Y (1992) Polym International 28: 81
48. Szeverenyi NM, Bax A, Maciel GE (1983) J Am Chem Soc 105: 2579
49. Caravatti P, Deli JA, Bodenhausen G, Ernst RR (1982) J Am Chem Soc 104: 5506
50. Cross TA, Frey MH, Opella SJ (1983) J Am Chem Soc 105: 7471
51. Linder M, Henrichs P, Hewitt JM, Massa DJ (1985) J Chem Phys 82: 1585
52. Packer KJ, Pope JM, Yeung RR, Cudby MEA (1984) J Polym Sci Polym Phys Ed 22: 589
53. Havens JR, VanderHart DL (1985) Macromolecules 18: 1663

54. Colquhoun IC, Packer K (1987) Brit Polym J 19: 151
55. Clauss J, Schmidt-Rohr K, Spiess HW (1993) Acta Polymer 44: 1
56. Schmidt-Rohr K, Clauss J, Spiess HW (1992) Macromolecules 25: 3273
57. Tekely P, Palmas P, Mutzenhardt P (1993) Macromolecules 26: 7363
58. (a) Cai WZ, Schmidt-Rohr K, Egger N, Gerharz B, Spiess HW (1993) Polymer 34: 267; (b) Spiegel S, Schmidt-Rohr K, Boeffel C, Spiess HW (1994) Polymer 34: 4566
59. Bailey RT, North AM, Pethrick RA (1981) Molecular motions in high polymers. Clarendon, Oxford
60. McCrum NG, Read BE, Williams G (1991) Anelastic and dielectric effects in polymeric solids. Dover, New York
61. Heijboer J (1960) Kolloid Z 171: 7
62. Sauer JA (1971) J Polym Sci Polym Symp 32: 69
63. (a) Spiess HW, Sillescu HJ (1981) Magn Reson 42: 381; (b) Spiess HW (1980) J Chem Phys 72: 6755; (c) Pschorn U, Rössler E, Sillescu H, Kaufmann S, Schaefer D, Spiess HW (1991) Macromolecules 24: 398; (d) Kaufmann S, Wefing S, Schaefer D, Spiess HW (1990) J Chem Phys 1: 197
64. Spiess HW (1984) Adv Polym Sci 66: 23
65. Axelson DE, Mandelkern L, Popli R, Mathieu PJ (1983) Polym Sci Polym Phys Ed 21: 2319
66. Kaji A, Yamanaki A, Murano M (1990) Polymer J 22: 893
67. (a) Sefcik MD, Schaefer J, May FL, Raucher D, Dub SM (1983) J Polym Sci, Polym Phys Ed. 21: 1041; (b) Sefcik MD, Schaefer J (1983) J Polym Sci Polym Phys Ed 21: 1055; (c) Sefcik MD, Schaefer J, Stejskal EO, McKay RA, Dixon WT, Cais RE (1984) Macromolecules 17: 1107
68. Liu Y, Roy AK, Jones AA, Inglefield PT, Ogden P (1990) Macromolecules 23: 968
69. Boyer RF (1968) Polym Eng Sci 8: 161
70. Schaefer J, Stejskal EO, Buchdahl R (1977) Macromolecules 10: 384
71. Shieh YT, Parker AA, Marcinko JJ, Ritchey WM (1990) Polym Prepr (Am Chem Soc Div Polym Chem) 31(1): 174
72. Roy AK, Jones AA, Inglefield PT (1986) Macromolecules 19: 1356
73. Cauley BJ, Cipriani C, Ellis K, Roy AK, Jones AA, Inglefield PT, McKinley BJ, Kambour RP (1991) Macromolecules 24: 403
74. Spiess HW (1983) Colloid & Polymer Sci 261, 193
75. Spiess HW (1983) J Mol Structure 111: 119
76. Inglefield PT, Amici RM, O'Gara JF, Hung C-C, Jones AA (1983) Macromolecules 16: 1552
77 Jones AA, O'Gara JF, Inglefield PT, Bendler JT, Yee AF, Ngai KL (1983) Macromolecules 16: 658
78. Hutnik M, Argon AS, Suter UW (1991) Macromolecules 24: 5970
79. Henrichs PM, Nicely VA, Fagerburg DR (1991) Macromolecules 24: 4033
80. Cholli AL, Dumais J, Engel A, Jelinski LW (1984) Macromolecules 17: 2399
81. Dumais JJ: Cholli AL, Jelinski LW, Hedrick JL, McGrath JE (1986) Macromolecules 19: 1884
82. Cain EJ, Gardner KH, Gabara V, Allen SR, English AD (1991) Macromolecules 24: 3721
83. Pschorn U, Spiess HW, Hisgen B, Ringsdorf H (1986) Makromol Chem 187: 2711
84. Heijboer JJ (1968) Polym Sci Part C 16: 3755
85. Schaefer D, Hansen M, Blümich B, Spiess HW (1991) J Non-Cryst Solids 131–133: 777
86. Flocke HA (1962) Kolloid Z 180: 118
87. Franck A, Biederbick K (1984) Kunstoff-Kompendium, 3rd ed. Vogel, Würzburg: p 277
88. Hagemeyer A, Schmidt-Rohr K, Spiess HW (1989) Adv Magn Reson 13: 85
89. Schaefer D, Spiess HW, Suter UW, Fleming WW (1990) Macromolecules 23: 3431
90. Wefing S, Kaufmann S, Spiess HW (1988) J Chem Phys 89: 1234
91. Zemke K (1994) Ph.D. Thesis, Mainz-Johannis Gutenberg Universität
92. Ernst RR, Bodenhausen G, Wokaun A (1987) Principles of magnetic resonance in one and two dimensions. Clarendon Press, Oxford: Chapter 9
93. Zemke K, Chmelka BF, Schmidt-Rohr K, Spiess HW (1991) Macromolecules 24: 6874
94. Flory PJ (1969) Statistical mechanics of chain molecules. Interscience, New York
95. Axelson DE (1986) In: Komoroski RA (ed) High resolution NMR spectroscopy of synthetic polymers in bulk. VCH, New York: Chapter 6
96. Fröhlich H (1942) Proc Phys Soc Lond 54: 422
97. Boyd RH (1985) Polymer 26: 1123
98. Schmidt-Rohr K (1991) Ph.D. Thesis, Mainz-Johannis Gutenberg Universität
99. Wilhelm M (1992) Diploma Thesis, Mainz-Johannis Gutenberg Universität: p 65

100. de Jong AF, Kentgens A, Veeman WS (1984) Chem Phys Lett 109: 337
101. Kentgens APM, de Jong AF, de Boer E, Veeman WS (1985) Macromolecules 18: 1045
102. Kentgens APM, de Boer E, Veeman WS (1987) J Chem Phys 109: 6859
103. Spiess HW, Schmidt-Rohr K (1992) Polym Prepr (Am Chem Soc Div Polym Chem) 33(1): 68
104. Schmidt-Rohr K, Spiess HW (1991) Macromolecules 24: 5288
105. Syi J-L, Mansfield ML (1988) Polymer 29: 987
106. Rutledge GC, Suter UW (1992) Macromolecules 25: 1546
107. Hoffmann JD, Williams G, Passaglia E (1966) J Polym Sci C14: 173
108. Opella SJ, Waugh JS (1977) J Chem Phys 66: 4919
109. Tonelli AE, Schilling FC (1981) Acc Chem Res 14: 233
110. Earl WL, VanderHart DL (1979) Macromolecules 12: 762
111. Schmidt-Rohr K, Kulik AS, Beckham HW, Ohlemacher A, Pawelzik U, Boeffel C, Spiess HW (1994) Macromolecules, accepted
112. Kulik AS, Spiess HW (1994) Makromol Chemie 195: 1755
113. Heijboer J (1982) In: Pethrick RA, Richards RW (eds) Static and dynamic properties of the polymeric solid state. D Reidel Publishing Company
114. McCrum NG, Read BE, Williams G (1991) Anelastic and dielectric effects in polymeric solids. Dover, New York: p 246
115. Kulik AS, Beckham HW, Schmidt-Rohr K, Radloff D, Pawelzik U, Boeffel C, Spiess HW (1994) Macromolecules, accepted
116. Leisen J, Boeffel C, Dong RY, Spiess HW (1993) Liquid Crystals 14: 215
117. Schwiegk S, Werth M, Leisen J, Wegner G, Spiess HW (1993) Acta Polym 44: 31
118. Chin YH, Zhang C, Wang P, Inglefield PT, Jones AA, Kambour RP, Bendler JT, White DM (1992) Macromolecules 25: 3031
119. Zhang C, Wang P, Jones AA, Inglefield PT, Kambour RP (1991) Macromolecules 24: 338
120. Shi J-F, Dickinson LC, MacKnight WJ, Chien JCW, Zhang C, Liu Y, Chin YH, Jones AA, Inglefield PT (1993) Macromolecules 26: 1008
121. Samuels RJ (1974) Structured polymer properties. Wiley, New York
122. Ward IM (1975) Structure and properties of oriented polymers. Wiley, New York
123. Kakudo M, Kasai N (1972) X-ray diffraction by polymers. Elsevier, New York
124. (a) Bower DI, Maddams WF (1989) The vibrational spectroscopy of polymers. Cambridge University Press, Cambridge: p 292; (b) Griffiths PR, de Haseth JA (1986) Fourier transform infrared spectrometry. Wiley, New York: p 437
125. Bower DI (1972) J Polym Sci Polym Phys Ed 10: 2135
126. Rodriguez F (1989) Principles of polymer systems, 3rd ed. Hemisphere Publishing Corp, New York: p 48
127. Bower DI (1981) J Polym Sci Polym Phys Ed 19: 93
128. McBrierty VJ, Ward IM (1968) Brit J Appl Phys (J Phys D) 1: 1529
129. Hentschel R, Sillescu H, Spiess HW (1981) Polymer 22: 1516
130. Röber S, Zachmann HG (1992) Polymer 33: 2061
131. Spiess HW (1981) In: Ward IM (ed) Developments in oriented polymers, vol 1. Applied Science Publishers, London: pp 47–78
132. Hentschel R, Schlitter J, Sillescu H, Spiess HW (1978) J Chem Phys 68(1): 56
133. Boeffel C, Spiess HW (1988) In: McArdle CB (ed) Side chain liquid crystal polymers. Blackie, Glasgow
134. Boeffel C, Spiess HW (1988) Macromolecules 21: 1626
135. Brandolini AJ, Apple T, Dybowski C, Pembleton RG (1982) Polymer 23: 39
136. Brandolini AJ, Alvey MD, Dybowski C (1983) J Polym Sci Polym Phys Ed 21: 2511
137. Hempel G, Schneider H (1982) Pure Appl Chem 54: 635
138. Garroway AN, Stalker DC, Mansfield P (1975) Polymer 16: 161
139. Henrichs PM (1987) Macromolecules 20: 2099
140. Schmidt-Rohr K, Hehn M, Schaefer D, Spiess HW (1992) J Chem Phys 97(4): 2247
141. Chmelka BF, Schmidt-Rohr K, Spiess HW (1993) Macromolecules 26: 2282
142. (a) Carter CM, Alderman DW, Grant DM (1985) J Magn Reson 65: 183; (b) Hughes CD, Sherwood MH, Alderman DW, Grant DM (1993) J Magn Reson 102(A): 58
143. Fina LJ, Koenig JL (1986) J Polym Sci Polym Phys Ed 24: 2509
144. Cullity BD (1978) Elements of X-ray diffraction, 2nd ed. Addison-Wesley, Reading, Massachusetts: p 295
145. Wilchinsky ZW (1963) Advances in X-ray analysis, vol 6. Plenum, New York: p 231

146. Harbison GS, Spiess HW (1986) Chem Phys Lett 124: 128
147. Harbison GS, Vogt V-D, Spiess HW (1987) J Chem Phys 86: 1206
148. Tzou D-L, Huang T-h, Abhiraman AS, Desai P (1994) J Polym Sci Polym Phys Ed, submitted
149. Vogt V-D, Dettenmaier M, Spiess HW, Pietralla M (1990) Colloid Polym Sci 268: 22
150. (a) Tzou D-L, Huang T-h, Abhiraman AS, Desai P (1992) Polymer 33: 426; (b) Tzou D-L, Schmidt-Rohr K, Spiess HW (1994) Polymer, accepted
151. Hong J, Harbison GS (1990) Colloids and Surfaces 45: 313
152. Féaux de Lacroix S, Titman JJ, Hagemeyer A, Spiess HW (1992) J Magn Reson 97: 435
153. Herzfeld J, Berger AE (1980) J Chem Phys 73: 6021
154. Titman JJ, Féaux de Lacroix S, Spiess HW (1993) J Chem Phys 98(5): 3816
155. Wilhelm M, Féaux de Lacroix S, Titman JJ, Schmidt-Rohr K, Spiess HW (1993) Acta Polymer 44: 279
156. Sanders JKM (1993) Chem Soc Rev 1
157. Kümmerlen J, Sebald A (1993) J Am Chem Soc 115, 1134
158. Titman J, Luz Z, Spiess HW (1992) J Am Chem Soc 114: 3756, 3765
159. (a) Yang Y, Hagemeyer A, Blümich B, Spiess HW (1988) Chem Phys Lett 150: 1; (b) Yang Y, Hagemeyer A, Spiess HW (1989) Macromolecules 22: 1004
160. Frydman L, Lee YK, Emsley L, Chingas GC, Pines A (1993) J Am Chem Soc 115: 4825
161. Hansen M, Blümich B, Boeffel C, Spiess HW (1992) Macromolecules 25: 5542
162. Günther E, Blümich B, Spiess HW (1992) Macromolecules 25: 3315
163. Mirau P, Heffner S, Koegler G, Bovey F (1991) Polymer International 26: 29
164. Bielecki A, Burum DP, Rice DM, Karasz F (1991) Macromolecules 24: 4820
165. Burum DP, Bielecki A (1991) J Magn Reson 94: 645
166. Kaszynski P, Friedli AC, Michl J (1992) J Am Chem Soc 114: 601
167. Tzou D-L, Spiess HW, Curran S (1994) J Polym Sci, accepted

# Author Index Volumes 21-32

MIX
Papier aus verantwortungsvollen Quellen
Paper from responsible sources
FSC® C105338
FSC
www.fsc.org

If you have any concerns about our products,
you can contact us on
**ProductSafety@springernature.com**

In case Publisher is established outside the EU,
the EU authorized representative is:
**Springer Nature Customer Service Center GmbH**
**Europaplatz 3, 69115 Heidelberg, Germany**

Printed by Libri Plureos GmbH
in Hamburg, Germany